Design Research: The Sociotechnical Aspects of Quality, Creativity, and Innovation

Dorian Marjanović · Mario Štorga · Stanko Škec
Editors

Design Research: The Sociotechnical Aspects of Quality, Creativity, and Innovation

Springer

Editors
Dorian Marjanović
Chair of Design and Product Development
University of Zagreb Faculty of Mechanical
Engineering and Naval Architecture
Zagreb, Croatia

Mario Štorga
Chair of Design and Product Development
University of Zagreb Faculty of Mechanical
Engineering and Naval Architecture
Zagreb, Croatia

Stanko Škec
Chair of Design and Product Development
University of Zagreb Faculty of Mechanical
Engineering and Naval Architecture
Zagreb, Croatia

ISBN 978-3-031-50487-7 ISBN 978-3-031-50488-4 (eBook)
https://doi.org/10.1007/978-3-031-50488-4

This Springer imprint is published by the registered company Springer Nature Switzerland AG
The registered company address is: Gewerbestrasse 11, 6330 Cham, Switzerland

Paper in this product is recyclable.

Preface

The essays collected present reflections of ideas, views, propositions, or alternatives presented and discussed at the workshop *The Design Research Sociotechnical aspects of Quality, Creativity, and Innovation* organised by the Chair of Design and Product Development at the University of Zagreb, Faculty of Mechanical Engineering and Naval Architecture.

The workshop participants represent engineering design and design science researchers from different universities and backgrounds. Most of the 40 participants of the workshop are networked through the Design Society, an interdisciplinary community of academics and industry practitioners interested in design research and practice. The Design Society fosters a multicultural, multidiscipline approach to design research insisting on scientific rigour.

The workshop's discussions were organised into three sessions: Evolution, Edification, and Challenge. The speakers of the Evolution session presented insights into how the design research has developed to the current stage and discussed the future design research prospects. A few educational projects were presented during the Edification session. The experience gathered in these projects, applied at several EU universities, provides valuable data and understanding for future design courses. The speakers at the Challenge session addressed key issues of design research: creativity, quality, human-centred approaches, and requirements of new technologies and smart products.

Through the discussions, participants highlighted the importance of sociotechnical aspects in engineering design. Besides functionality and other technical aspects, design involves the responsibility for sociotechnical aspects such as quality, creativity, and innovation. These aspects are interrelated and can significantly impact the success of an engineering design project.

Following the workshop discussions, the authors prepared contributions included in this volume. The book aims to provide insight and understanding of engineering design research, offering an articulated perspective on the development and enhancement of high-quality products, creativity, innovation, invention, and productivity.

We are thankful to all the workshop participants who were able to participate and contribute to discussions at the workshop. Further, we are thankful to all the authors who prepared the contributions collected in this book.

Zagreb, Croatia

Dorian Marjanović
Mario Štorga
Stanko Škec

Contents

Chapter 1
Introduction

Dorian Marjanović, Mario Štorga, and Stanko Škec

Abstract Engineering design processes have undergone transformative changes over time, driven by a confluence of factors that have enhanced their efficacy and pertinence. This evolution is intricately tied to technological advancements, shifting societal needs and values. In response to the dynamic needs and perspectives within the industry, design research has evolved into a user-centred, data-driven, and socially responsible field closely aligned with technological advancements. This shift has led to fundamental principles in design research, emphasizing quality, innovation, and creativity. Initially confined to engineering design, the members of The Design Society broadened their research scope into trans-disciplinary realms, showcasing the interdisciplinary nature of modern design research. This transformation was evident in a workshop in Zagreb, where participants reflected on the triad of quality, creativity, and innovation in design research. The introduction chapter of this book sets a scene, bringing a brief overview of the developments through time.

Over time, engineering design processes and practices have undergone remarkable transformations, fuelled by many factors that have enhanced their effectiveness and relevance. Technological advancements, evolving societal needs and values, and accumulating knowledge and experience have all played pivotal roles in this evolution.

Accordingly, designers have continually honed their skills and techniques, embracing a multi-faceted approach to improvement. They relied on trial-and-error methods in the early stages, drawing insights from their experiences. Subsequently, the introduction of formal education improved their expertise, providing them with a solid foundation to build upon. Moreover, ground-breaking technologies and scientific breakthroughs further propelled their progress.

This ongoing development in design practices has empowered designers to fashion products that are more sophisticated, innovative, and sustainable. By integrating a range of influences, this progress has culminated in the creation of solutions that

D. Marjanović (✉) · M. Štorga · S. Škec
University of Zagreb Faculty of Mechanical Engineering and Naval Architecture, Zagreb, Croatia
e-mail: dorian@fsb.hr

D. Marjanović et al. (eds.), *Design Research: The Sociotechnical Aspects of Quality, Creativity, and Innovation*, https://doi.org/10.1007/978-3-031-50488-4_1

harmonize with societal needs while simultaneously minimizing their environmental impact. The confluence of diverse factors has been instrumental in shaping this transformative journey.

The perception and understanding of design are also influenced and distorted by various factors such as media, policy, and domain-specific viewpoints. These external influences shape the perspectives on design, often beyond the design's actual purpose and value. Therefore, the design community has a crucial role and responsibility in fostering scholarly communication and discourse in academia and practice. Designers as practitioners and design researchers play a significant role in shaping our world, as design involves creating artefacts that we use, discard, and eventually decompose or extinguish. Therefore, the design community needs to engage in meaningful and open discussions about design, its impact, and its potential consequences. Design is omnipresent activity, creating artefacts that shape our daily lives. This activity emphasizes the significance of design in influencing our environment and culture and the need for a thoughtful and informed approach to design discourse.

Design community and practice must promote meaningful and unbiased discourse on design, highlighting the ubiquitous nature of design in shaping our world. The design discussions should be free from interests or biases and focused on the intellectual exchange of ideas, concepts, and perspectives to promote a holistic understanding of design and its implications.

The introduction of formal engineering education in universities has significantly contributed to improving designers' skills. The first engineering school was founded in France in the late eighteenth century. Since then, many universities worldwide have started offering engineering degrees, providing students with a strong foundation in design principles.

Engineering design education was predominantly influenced by scientific knowledge. However, as our understanding of science and engineering has grown, designers have had access to more information about how things work and how to design them effectively. This knowledge has been accumulated over time, with each generation building on the work of those who came before them. Franz Reuleaux, nowadays most known as a founder of modern kinematics, presented the first ideas of the academic approach to design. In addition to the theory of machines and machine design (Reuleaux, 1875) presented in "*A handbook of machine design*", his book *Der Constructeur*, printed in German in (Reuleaux, 1861) (the English translation of the book: *The Technology and Civilisation* (Reuleaux, 1891)), initiated a philosophical approach to machine knowledge (Maschinewessens) drawing connections between engineering and humanities by analysing the social implications of industrialisation.

The development of new technologies provided new tools, materials, production processes, and software has revolutionized how designers work—enabling designers to create more complex and sophisticated products. For example, advancements such as computer-aided design (CAD), rapid prototyping, and simulation tools have made it easier for designers to visualize their designs, test them, and refine them before creating physical prototypes.

Two additional factors, collaboration and integration, influenced the design practice. Collaboration once limited within departments, the development and production teams, evolved with the rise of global communication networks and the Internet. As a result, designers now collaborate with other designers and experts worldwide, allowing them to share knowledge, insights, and ideas that can improve their designs. In addition, as technology advances, designers have increasingly incorporated digital tools into the design process. For example, designers may use data analytics tools to analyse user behaviour patterns and identify areas for improvement.

Design research has developed over time in response to changing needs and perspectives within the design industry. The researchers and practitioners are focused on the particular topics of the design process or design problems considered. This way of thinking is motivated primarily by the core nature of engineering design, considered as a "process in which technologies materialise into products" (Kroes et al., 2008). The new products have a circular influence visible not only on technology; therefore, design can be perceived "as a process that substantively shapes and reshapes our lives and societies" (Kroes et al., 2008). It might be said that social implications of design are embedded throughout the whole design process, from early ideas to the artefact embodiment, production, distribution, usage and decomposition. Dixon (Dixon, 1966) positioned design on the crossroad of social-cultural and science-technical paths.

Such a development enabled engineers to create increasingly complex and sophisticated systems. As the world faces pressing challenges such as climate change and resource depletion, the skills of engineers and designers will be more critical than ever in creating a sustainable future for humanity.

Responding to the changing needs and perspectives within the design field, design research has evolved to become a more user-centred, data-driven, and socially responsible field closely integrated with technological advancements. Several views and research streams influenced such a development. The user-centred design approach emerged in the 1980s and has since become a dominant paradigm in design research. It emphasizes understanding users' needs and behaviours through research and observation.

As design research has shifted toward user-centred approaches, qualitative research methods have become increasingly important. These methods, such as ethnography and user interviews, help designers gain a deep understanding of users' needs and behaviours.

The emergence of design thinking methodology positioned the user at the centre of the design process. It emerged in the 1980s and 1990s and has since become an aspect of design research and practice far beyond the design field.

Another multidisciplinary aspect had a significant influence. Research on creativity has had a significant impact on design research, as it has provided valuable insights into the creative processes and strategies that designers use to generate novel and innovative design solutions. Creativity research has explored various aspects of the creative process, including ideation, problem-solving, and the role of motivation and inspiration. This research influenced design research and practice, particularly in understanding the creative process, generating ideas, and providing various methods such as brainstorming or mind mapping. Understanding the creative process and the flow enabled design researchers and practitioners to incorporate creativity techniques to foster a more creative and innovative design process.

The quality and rigor in design research have become increasingly significant, particularly over the last two decades, owing to the recognition of their importance in shaping the quality of products, fostering innovation, and enhancing competitiveness. Design research achievements are reported in several conferences and journals. However, a clear picture of the matters that are design research problems and those which are problems of other research disciplines is still missing. Blessing describes design research as "aims at increasing our understanding of the phenomena of design in all its complexity, and at the development and validation of knowledge, methods, and tools to improve the current situation in design" (Blessing, 2002).

At the Workshop in Zagreb participants considered three aspects of design research, namely: quality, creativity, and innovation. Quality here refers to the degree of excellence of design research. Quality is critical in engineering design because it directly affects the product's ability to meet customer needs, comply with industry standards, and compete in the market. It encompasses various aspects such as reliability, durability, performance, and user satisfaction. Achieving high quality requires a systematic approach involving quality management tools and techniques.

Creativity is the ability to generate new ideas and approaches to solve problems. In engineering design, creativity is essential to develop innovative solutions that address complex and evolving customer needs. It is also a crucial component of the engineering design process; without it, there is no potential for innovation (Howard et al. 2006, 2010). Creativity can be fostered by encouraging experimentation, risk-taking, and cross-functional collaboration. The research on creativity in design aims to provide a better understanding of the processes that lead to creative designs and proposes integrating a creative process with the overall design process better to utilize creativity tools, methods, and techniques. Engineers must also possess a broad range of skills, including critical thinking, communication, and empathy, to facilitate the creative process (Cully et al., 2002). Creativity and innovation in engineering design are intrinsically linked, as creativity is the foundation for generating new ideas and approaches.

Innovation is successfully implementing a new idea or process that creates value. In engineering design, innovation is critical to developing competitive, sustainable products that meet customer needs. Innovation can take many forms, including incremental product improvements or radical breakthroughs introducing new technologies or business models. Successful innovation requires a culture that supports experimentation, embraces diversity, and rewards risk-taking.

The socio-technical aspects of quality, creativity, and innovation are intertwined and impact each other. A focus on quality can lead in both design research and design to more innovative solutions that improve the product's reliability and performance. Likewise, a culture of creativity can foster innovation and lead to products that meet customer needs more effectively. Engineering design teams must balance these aspects to achieve a successful outcome that meets customer needs, complies with industry standards, and is competitive in the market.

1.1 The Evolution of Design Research, a Path to Transdisciplinarity

The nature of product development has undergone a significant transformation in a relatively short period of time. The design focus has shifted from engineering design of mechanical systems to encompass technical artifacts, product service systems, usage activities, and symbolic representations. Nowadays, designers and product development teams are responsible for a product's entire lifespan, incorporating services and designing for re-use and pre-use patterns. This approach has resulted in significantly reduced response times for companies. Consequently, the research scope has broadened considerably over time.

The design research development has passed through four overlapping phases according to Wallace and Blessing: Experiential, Intellectual, Experimental (Wallace & Blessing, 2000) and Theoretical phase (Blessing, 2002) (Chakrabarty & Blessing 2014). The research goal in the first phase of research was mainly oriented to support the design practice, while the later phases considered support for research and practice.

The brief sketch of the design research presented here serves as a foundational context for the workshop topics, providing a general understanding of the subject matter. A comprehensive overview of the international design research and teaching outcomes can be found in the proceedings of the Design Society conferences (www.designconference.org). In addition, several reports provide a deep and detailed description of the matter and suggest future research challenges (Archer, 1981), (Hubka & Eder, 1987), (Finger & Dixon, 1989a, 1989b), (Cross, 1993), (Blessing, 2002), (Friedman, 2003), (Bayazit, 2004), (Horváth, 2004), (Birkhofer, 2011), (Andreasen, 2011), (Gericke and Blessing, 2012), (Cash, 2020) or (Krause & Heyden, 2022), to mention just a few in chronological order. Chakrabarti and Blessing (2014) prepared a very exhaustive, recent review of design research under the title "An Anthology of Theories and Models of Design—Philosophy, Approaches and Empirical Explorations".

The inauguration of design research accompanied the industrial revolution at the end of the eighteenth century. The earliest ideas have been related to machine design (Prof. Redtenbacher from Karlsruhe and his student Reuleaux). The beginnings of design research were initiated by the systematic design of machines. The first step-by-step approach in systematic design might be attributed to Erkens (1928), who presented a step-by-step approach in design. The approaches of the early and mid-twentieth were inherently connected with the development of technology. Wögerbauer presented one of the earliest approaches to systematic design by a hierarchical task structure distinguishing operational and implementational tasks (Wögerbauer, 1943).

The early research in engineering design was scattered by-product domains, language and cultural borders and the national organisation of professional societies of engineers and academics. The communication means, language and national borders prevented the cooperation of the researchers until the second part of the previous century. Quite a lot of early continental research was virtually unknown outside national borders. By the end of the 1970s, many individuals and research groups were engaged in design research with little interaction due to cultural, language and design terminology differences that hindered common understanding or discussion (Hales 2005).

The first design models inherently arose from the researcher's experience and knowledge gathered over engineering design practice and drilled through the development of complex engineering products. In this experiential phase of research (Wallace & Blessing, 2000), researchers rationalised their own processes. Finally, the gathered knowledge was extrapolated and generalised, like in the work of Kesselring (1942).

Kesselring laid down the principles of minimalization concerning production costs, weight, space requirements, losses and handling (Kesselring, 1951, 1954, 1955). The context of design research at that time was focused on the design of machine elements like in the works of (Matousek, 1957) and (Niemann, 1975), or on general engineering design methodology like in the works of (Roth, 1982), and (Koller, 1985). The generations of mechanical engineering students were trained with Niemann's textbooks in machine design. Niemann insisted on task definition and refinement followed by systematic variation of possible solutions and selection of the optimum solution.

Rodenacker (1970), Rodenacker and Claussen, 1973, 1975 developed a comprehensive design method that focuses on establishing the working interrelationship of a technical system through logical, physical, and embodiment relationships. This approach highlights the early identification and mitigation of disturbances and failures by adopting a selection strategy and evaluating parameters against multiple criteria. A key element of Rodenacker's method is establishing the physical process, which enables a systematic search for innovative solutions. The utilization of morphological analysis, derived from the work of astronomer Zwicky (1966), serves as a method for identifying potential problem solutions based on a function structure diagram.

The effort of the researchers resulted in a set of procedural models and design recommendations known as Methodical Design (Methodische Konstruiren). The most known researchers who contributed to design methodology, in addition to already mentioned names, are Frankenberger, Koller, Ehrlenspiel, Pahl and Beitz. The design methodology in the research efforts mentioned was based on mechanical engineering design. Design Methodology procedures, including checklist tools and design catalogues (Roth, 1982), were a combination of pragmatic, experience-based tools and physical principles. It should empower designers to foster a procedural approach, thus enabling novice designers to systematically approach the problem, ideate several solution-neutral design proposals, evaluate and develop solutions to complex design problems. The book "Konstruktionslehre" (Pahl & Beitz 1977) became a reference textbook for systematic design. The first English version, entitled Engineering Design (Pahl & Beitz, 1984), was translated by Wallace and Blessing and published in 1984. This effort, the translation of such a fundamental work, had a tremendous influence on the internationalisation of design research outside of the german speaking area. The third edition of the English translation is published under the title Engineering Design: a systematic approach (Pahl et al., 2007).

In the mid of the previous century, design research was still merely separated by engineering domains and national borders and concentrated on design methods. Similar efforts could be traced in other countries forming separate lines of research in Design Methodology, like those conducted in the former German Democratic Republic by Hansen in 1950, (Bischoff and Hansen 1953), (Hansen, 1966) or in Poland by Dietrich (Hubka, 1983).

The Ilmenau School (Bischoff and Hansen, 1953), proposed a systematic design approach in the 1950s, and Hansen presented a more comprehensive approach in 1966 (Hansen 1966, 1974). The approach involves four working steps, starting with analysing and specifying the task, then a systematic search for solution elements, analysing and improving the working means, and evaluating the optimum working means.

The establishment of the chair on engineering design at DTU by Jeppesen in 1952 sparked design research in Scandinavia (Alger & Hays, 1964). Tjalve established the foundations of industrial design based on Hubka's Theory of Technical Systems (TTS) (Andreasen & McAloone, 2008). Tjalve promoted, for industrial design standards of the time novel approach, that form should follow function (Tjalve, 1976). The work on Design for Assembly (DFA), based on Hubka's concept of DfX (Hubka, 1974), was motivated by intense cooperation with industry. In the late 70 s, DTU focused on assembly automation, leading to the publication of "Design for Assembly" in 1982 and "Flexible Assembly Systems" in 1986. Olensen, 1992 expanded on Tjalve's "fitness for life" concept by viewing each life phase as a transformation system with unique characteristics that influence the product's optimal performance. Stakeholder criteria for optimal performance must balance cost, quality, time, efficiency, flexibility, risk, and environmental effects. Olesen created a DFX Matrix, which encompasses DFX areas. According to (Andreasen & McAloone 2008), Olesen's Dispositions theory can be considered a comprehensive theory that covers a variety of product life and universal virtue concerns within this matrix.

The work of J. Dietrych in Poland was virtually unknown until ICED conferences in Rome in 1981 and Copenhagen in 1983. Dietrych proposed "General concepts of design science", elaborating his approach in his work "Reasons and principles—the fundamental of the design theory" at the 3rd World IFTOM conference in 1971. Dietrych presented a System of design tasks as a networked approach comprising of activities (nodes) and transformations (branches) in two parallel domains: abstract and concrete (Hubka, 1983).

In the UK, the design research was scattered into engineering and industrial design. (Wallace, 1952) considered the technique of engineering design, discussing the nature and origins of design and highlighting the responsibility of designers. The importance of design in mechanical engineering was stressed in a Feilden report (Feilden, 1963). This report had a substantial influence on engineering design education in Great Britain. In addition, the emergence of new disciplines like operational research, organisation and management, decision-making, creative techniques, and computer programming influenced the authors of new design methods. Archer (1954) argued the importance of artistic and engineering components in product design during the fifties. He presented the Systematic Method, expressing a holistic view on design in a sequence of articles published in Design magazine during 1963–64 (Archer, 1963a, 1963b, 1963c, 1963d, 1964a, 1964b, 1964c). The magazine was published by Design Council, established in 1949 as Industrial Design Council. In the proposed design process model, distinct approaches were necessary for various stages of industrial design. During the analytical phase, systematic observation and inductive reasoning are required, while subjective and deductive reasoning are needed in the creative phase.

Archer's method, constructed on scientific practices, triggered discussion in the field, arguing that the prescribed method obliged designers to act as machines (Jones, 1966). Design Research Society reprinted Archer's work with a valuable review of publications and work (Lloyd et al., 2016), (Boyd Davis and Gristwood 2016) on the occasion of the 50th anniversary of the Design Research Society. The general principles of scientific research for obtaining data for engineering design were in the 1970s elaborated by (Glegg, 1973).

In the US, several authors like (Gordon, 1961), (Osborn, 1963), and (Dixon, 1966) focused their work on design methods that would lead to creativity or innovation. The books "Introduction to Design" (Asimov, 1962) and "The Sciences of the Artificial" (Simon, 1969) laid down a broader, multidisciplinary, and scientific approach to design research, bringing, besides classical engineering subjects, the cognitive sciences and economics into consideration. Finally, in the third edition of "The Science of the Artificial", classic in the design science literature, Simon discussed the emerging topics of the time, like chaos or genetic algorithms used in the analysis of complex systems.

The design methodology was developed to increase effectivity and efficiency in the design office, distinguish artistic approach to design and "... enable appropriate, controlled and verifiable procedure to obtain resilient solutions ..." (Pahl et al., 2007). In addition, the methods were developed to foster the intuition and individuality of the designers. The continental, mainly German research, highlighted systematic methods and approaches to design problems, while research in the USA fostersed creative phases of the design process (Birkhofer, 2011). The researchers in UK (French, 1985), Cross, 2006) and USA had a product-oriented approach to concept generation, whereby the analysis of an initial product idea is the primary focus. This idea is then refined step-by-step to create a concrete concept without much support for the generation process beyond generic idea generation methods.

The primary means of disseminating research outcomes in design research during that time were textbooks, professional journals published by national engineering associations, and national (engineering) conferences. Textbooks provided a comprehensive and structured format for presenting theoretical frameworks and research findings. They served as valuable resources for students, researchers, and professionals, offering in-depth explanations, case studies, and practical applications. Professional journals published by national engineering associations were important platforms for practitioners and researchers to share their work with the broader community. They provided a means to extend the body of knowledge in design practice and research, enable feedback from experts in the field, and engage in scholarly discussions. Textbooks and journal publications often underwent rigorous editorial processes to ensure accuracy and credibility.

National engineering conferences offered a forum for researchers, practitioners, and industry professionals to come together and exchange ideas. These conferences typically featured presentations, panel discussions, and workshops on various design research topics. Presenting research outcomes at conferences allowed researchers to receive immediate feedback, gain visibility within the community, and establish collaborations with other experts in the field.

The conference on Design methods held in London in 1962 was, in that sense, a milestone, a significant gathering of professionals and experts from diverse fields like industrial design, engineering, architecture, and communications. The main objective of this conference was to provide a platform for individuals and groups who share common interests and purposes to come together and discuss their ideas, methods, and experiences regarding design issues. The conference organizers aimed to explore the application of scientific methods and knowledge to solve the specific problems faced by individuals and groups in their respective fields. Moreover, the organizers hoped to promote interdisciplinary collaboration and break down the barriers between different creative activities. The conference attendees focused on discussing systematic and intuitive methods in various fields and how they could be integrated to create more innovative and effective solutions. As a result, the conference helped establish new connections between different fields, leading to a better understanding of the potential for cross-disciplinary collaboration in the creative industries of the time.

The seventies and eighties brought computers into design and the internationalization of research. New tools, technologies and communication means emerged in a short period, the design research became widely accessible, and the first attempts at consolidation began to appear. The research moved the focus from product design to product development. The complexity of the new products emerged manufacturing technologies, and product development performed in globally distributed conditions created a new scene and new demands for design research. New design research opportunities were created nationally, like programmes for design research established by the National Science Foundation (NSF) in the USA and the UK's Science and Engineering Research Council (SERC). The new journals devoted to design research were to appear (e.g. Research in Engineering Design and the Journal of Engineering Design).

For the internationalisation of design research, the work of Vladimir Hubka (Hubka, 1980, 1982), (Hubka and Eder, 1992, 1996) was very significant. With the help of professors Mogens Myrup Andreasen and Umberto Pighini, Hubka established an informal group, WDK (Workshop Design-Konstruktion) in 1978, based on a common interest in engineering design, insisting on internationality even in the title of the group. The group organised the first ICED conference in 1981 in Rome. The stage of conference events was also changed with other events considering design like the CIRP Design seminar, ASME DTM (Design Theory and Methodology) conference, DESIGN conference and others. Finally, in the year 2000, after 15

ICED conferences and more than 30 WDK publications, including a four-language dictionary of design research terminology, the activities of WDK were transferred to The Design Society.

Computers in design were introduced in various stages of the design process and for different purposes ranging from physical state analysis, process support, and new research themes focused on the geometric representations in CAM and CAD. In addition to the development of CAD/CAM/CAE software tools, several attempts were oriented to develop an software environment or tool to support a design process based on design methods. Programs were developed to design by parametrised parts or forms like in the work of (Claussen, 1971). Several research groups simultaneously proposed the CAx support based on design methodologies. Designers' workbench (Andreasen & McAloone 2008) was an attempt to utilise Domain theory (Andreasen et al., 2014). Another approach was based on network models of the design process like (Marjanovic & Pavkovic 1995) (Marjanovic 1996). Cully and McMahon considered the design process as an information processing activity (McMahon et al. 2005), (Cully, 2014).

The complexity was addressed in the attempts to develop design theories by many researchers, besides of already mentioned Hubka's Theory of Technical Systems (TTS) (Hubka and Eder 1974, Hubka & Eder, 1988). New theories have been proposed building upon the existing theories technical or systemic theories, like Andreasen's Domain Theory (Andreasen and McAloone 2008, Andreasen, 2011), Andreasen et al., (2014) or independently, like Yoshikawa's General Design Theory (Yoshikawa, 1981) and Extended General Design Theory (Tomiyama and Yoshikawa 1986). Such a holistic approach (Birkhofer, 2011) broadened the research scope. Roozenburg and Eeckels extended the scope of the design process from product design projects to product planning (Roozenburg and Eeckels 1995), highlighting the need to open the problem space, thus enabling innovation. Gero introduced the cognitive domain in Function-Behaviour-Structure (FBS) ontology (Gero, 1990).

Ehrlenspiel (Ehrlenspiel, 1985) emphasized the significance of management in the design process and product development. Within the context of systematic design, Ehrlenspiel introduced target costing and cost-driven design as essential components.

The "Characteristics-Properties Modelling" (CPM) and "Property-Driven Development" CPM/PDD approach proposed by (Weber, 2005) was developed in the late 1990s to study designers' handling of methodologies and methods. CPM represents the product model through characteristics and properties, and PDD depicts the process of developing and designing products based on CPM. The approach distinguishes between characteristics and properties of a product, with characteristics being directly determined and properties being indirectly influenced by modifying the characteristics. The approach uses relations, dependencies, and external conditions to represent constraints and context. Weber distinguished four steps in the PDD process: synthesis, analysis, individual deviation, and overall evaluation, which are carried out iteratively until all properties fulfil the required properties sufficiently, and all necessary characteristics needed for manufacturing and assembly of the product are established and assigned.(Weber, 2005, 2008).

Albers proposed the C&CM (Component & Connectivity Modeling) method that integrates technical functions and product shape into a single model (Albers et al., 2008). It allows designers to link abstract functions to product geometry, enabling them to generate, modify, or evaluate designs in an integrated way. The core of C&CM is a systematic function-component mapping that identifies and visualizes functions within components.

The complexity and wideness of design research were also addressed by the researchers at the Institute of Product Development at TUM. The Institute made significant contributions to design research in several areas, including methodical design of technical systems (Lindemann, 2006), innovative product development, concept development (Ponn & Lindemann, 2011), bio-inspired design (Farzaneh & Lindemann 2011), complexity management (Lindemann, 2009, Ponn & Lindemann, 2011), and human behaviour in design (Lindemann, 1998). These topics were particularly relevant for industrial applications, where complexity management is crucial for successful product development (Weber and Birkhofer 2007), (Chakrabarty and Lindemann 2015).

Between 1980 and 2000, there was a surge in the design research field and its outcomes. New factors such as knowledge, user experience, use scenarios and business strategy were introduced. This surge was driven by the growing recognition of the importance of design in various fields, such as product development, architecture, and urban planning and the increasing complexity of the design challenges faced by practitioners. After 2000, several formal theories were proposed: Axiomatic Design presenting a matrix-based methodology to analyse and transform user needs into functional requirements, design parameters and process variables (Suh, 2001), C-K design theory or, Concept-KnowledgeTheory (Hatchuel and Weil 2003). In addition, new approaches, methods and tools have been proposed. Design Thinking (Brown 2008) the Design Structure Matrix and focus on product development (Ulrich and Eppinger 2004) gained attention in recent years as companies and organisations seek to improve their design processes and create more effective products and services. These methods represent a shift towards more collaborative, user-centred approaches to design and are often seen as an alternative to more traditional, top-down design methods. Such development led to knowledge and information management topics in design research. Culley studied and classified the information generated and used during the design process (Culley, 2014; Culley et al., 2002). Information management becomes crucial since the design process's linear progress (regardless of planned or unplanned iterations and feedback in some early models) has evolved into parallel evolution in different spaces.

The network analysis (Petric Maretic, 2014) of research reports presented at the DESIGN conference in 2002–2014 has shown how the topics in design research have changed over the years. New fields such as psychology, cognitive sciences, engineering systems and others have been included in the research. This result is in line with the work of Friedman (2003), in which research areas that contribute to design research are identified as natural sciences, humanities and liberal arts, social and behavioural sciences, human professions and services, creative and applied arts, in addition to technology and engineering. In recent years, there has been a growing emphasis on inclusion (Keates & Clarkson, 2003), sustainability and social responsibility in design research. Designers are increasingly focused on creating products and services that are environmentally friendly and contribute positively to society. An extensive analysis of the challenges facing design research, particularly in terms of research impact and theory development is reported by Cash (2020). In this work Cash proposed a theoretical "landscape" of design research. He identified the opportunities for theory development related to: Design process, methods and tools, and Team processes. Cash also calls for multidisciplinary approach in design research integrating with "1: Cognition, reasoning, and behaviour; 2: User and social system; 3: Managerial and organisational system".

The evolution of design research has encompassed a shift from a relatively narrow focus on machine design, broadening interest to product development, up to a wider and more comprehensive examination of product-service systems and their societal ramifications. This paradigm shift has resulted in several fundamental tenets of design research: the emphasis on quality, innovation, and creativity. Once limited to the engineering design domain, the research expanded to trans-disciplinary research. Such expansion of research beyond the engineering design domain highlights the interdisciplinary nature of contemporary design research. Presented contributions indicate an overview of the collective body of work, suggesting a comprehensive understanding of the field. This can be recognised as a focal point in multiple contributions in this volume.

1.2 Overview of the Contributions

The first part of this volume offers three different views on the evolution of design research. Although the research is inherently connected with the development of technology and understanding of design thinking, the current knowledge of design in all the broadness incorporates technical, social and policy attributes.

The development of the design research community over four decades, from 1981 to the present, has been a story of growth, consolidation, and diversification. Nevertheless, addressing the societal challenges in a coherent, systematic, structured way is a fundamental strength of design: research, methods and practice, rising from a deep understanding of the needs, values, and aspirations of people and the ability to translate this knowledge into innovative products, services, and systems that address societal problems.

In the chapter: **"From Design Optimization to Design Science: An Evolution in Design Thinking"**, Panos Y. Papalambros derives interconnected perspectives often used in engineering design to approach problem-solving, innovation, and decision-making in a systematic and holistic manner.

The concept of design as a decision-making process has been widely used in engineering design, allowing designers and stakeholders to choose alternatives throughout the design process. In addition, this paradigm has enabled the use of mathematical formulations in design. Finally, it has led to the recognition of design optimization as a major area of research and practice.

Elucidating the relationship between the evolution of Design Optimization, Design Thinking, and Design Science author reflects on the evolving nature of design as a multidimensional and interdisciplinary field. Furthermore, design is viewed not solely as a technical or artistic endeavour but also as a social, cultural, and philosophical one. Integrating these three aspects leads to a more comprehensive and balanced approach to design, incorporating quantitative analysis, human-centred perspective, and holistic understanding, ultimately contributing to more sustainable, responsible, and meaningful design outcomes.

The chapter: **"DESIGN, Building a Global, Interdisciplinary Community"**, written by Chris McMahon and Anja Maier, enlightens the development of design research shaped by the political and economic Zeitgeist of the times. In the 1980s, the design community was emerging, and design research became mainstream. It was an era of significant scientific and cultural development, with strong achievements in the performing arts, but it was undoubtedly politically traumatic. There was stagflation and the fall of the Bretton Woods system of international monetary management, all associated with the oil crisis of the time. On the other hand, the 1980s were a major turning point for computing, and these developments benefited computer-aided design (CAD) owing to the introduction of low-cost graphics capabilities for gaming. In the 1990s, the design community experienced substantial growth, manifested in the increased number of conferences, journals, and engineering design research centres established worldwide. In the 2000s, the community went through a period of consolidation, with the development of design research in many new fields, such as health, energy, transportation, agriculture, manufacturing, and policy.

Over time, design research has influenced numerous aspects of societal needs, such as the development of new medical technologies, improvements in healthcare delivery, advancements in renewable energy systems, the creation of safer, more efficient, and more sustainable modes of transportation, enhancements in production processes, and the development of new materials.

The professional conferences played a significant role in the development and growth of the design research community over time. Such gatherings contributed to building and organising groups of researchers in informal, like the former Workshop-Design-Konstruktion—WDK group, or formal societies like The Design Society. Conferences provide a platform for interdisciplinary discussions and research on the process of designing. In addition, the conferences contribute to the quality of design research implying standards of article reviewing. Through the conference topics, such events also shape the design research and practice.

In the chapter: **"Designing a design conference in an emerging design science community: Danish experiences from the International Conference on Design (DESIGN)"**, Tim C. McAloone and Mogens Myrup Andreasen present the development of the research community and research rigour through the case of DESIGN conference events and the impacts on the research groups.

The DESIGN conference is a long-standing event that predates the establishment of the Design Society, which was officially founded in 2000. However, the conference has been closely linked to Society since its early days, with collaborations between Zagreb University and the WDK (Workshop Design-Konstruktion) forming from the second event 1984. This close collaboration has continued to this day, with DESIGN leading the way regarding article reviewing standards and integration of workshops into the conference programme.

According to the authors, the DESIGN conference has significantly impacted the community with its year-on-year improvements to article reviewing and careful matching of articles to reviewers' research areas leading to a high hit rate and quality of reviews. Additionally, the conference has shaped the topics discussed within the Design Society over the years, with a shift towards higher-level design management and diversely represented fields. The emergence of 'design research' as a topic at the conference reflects a focus on understanding the design process through research. The broadening of the research topics attracted researchers from various fields, including industrial design, anthropology, applied psychology, systems engineering, and design management. This broadening of the conference's scope has been a unique event characteristic, allowing for interdisciplinary discussions and diverse perspectives.

The second part of the book deals with design education. Two chapters present design educators' experiences, findings, and proposals. In addition, both chapters discuss the Problem-Based Learning (PBL) design courses realized in a demanding setting involving a multicultural and multidisciplinary collaboration of student teams from different universities and backgrounds.

Teaching, learning, and developing engineering design and design research skills are research topics for themselves. The emphasis has shifted from getting a basic knowledge about design to developing design problem-solving capabilities and improving thinking that will pursue creativity so critical for design. Design edification methods and approaches are broad, as are the design disciplines. Problem-based learning courses have become a popular educational approach in the engineering design domain, as they offer several educational benefits, such as increased student engagement and participation, communication skills, critical thinking, and industrial and real-life relevance of student assignments. This pedagogical approach emphasizes learning by doing, putting students in front of real or realistic project tasks, often complex and ill-structured by their own definition. Design projects within PBL courses are often organized around a central problem that drives students towards a tangible outcome and product.

PBL courses typically include working in smaller teams to collaboratively explore and understand the problem, analyze existing and propose new solutions, and evaluate and refine the proposed solution. Therefore, PBL facilitates collaborative learning and provides means to introduce a social-constructivist paradigm in a learning process. However, to maximize the benefits of PBL courses, educators must establish working structures and provide an environment where students with different backgrounds and knowledge can collaborate, both in online and traditional classroom settings.

Collaborative teaching, creativity, and innovation are critical in engineering and product design education. The defining and enabling skills that form the core competencies of the engineering graduate can be summarized in three roles: the engineer as a specialist, the engineer as an integrator, and the critical role of engineering graduates in providing the creativity, innovation, and leadership needed to guide the industry to a successful future. Many distinctive views on the development of these competencies can be identified. Two views, specifically, illustrate the wideness of the spectrum: the reductionist view, which assumes that design competence is nothing other than a set of fundamental design abilities addressed individually, and the holistic view, which sees design competency as a synergetic construct of generic human capacities.

Individual reflection and self-learning are essential aspects of PBL courses in design education, but students must also learn to collaborate and work within a team context. The interplay between individual and team activities within PBL courses is a critical aspect that has not been fully explored. To understand the strengths and weaknesses of ICT (Information and Communication Technology) systems, educators and researchers in engineering education and professionals and developers of Information and Communication Technology (ICT) tools for designing need to examine how the use of ICT tools changes during the design process. Understanding the interplay between individual and collaborative use of ICT tools in PBL courses is essential for providing structured support to students during the design of PBL courses, identifying potential pitfalls for collaborative design projects in remote settings, and developing ICT tools that meet the demands of geographically distributed design teams. However, the dynamics and interplay between individual and collaborative use of ICT tools during a design project-based learning course have not been extensively studied.

In the first chapter regarding education: **"Dynamics of Using Information and Communication Technology Tools in a Distributed Project-Based Design Course"**, Nikola Horvat, Niccolò Becattini, Harshika Singh, and Stanko Škec present the dynamics and interplay of individual and collaborative use of ICT tools during a design project-based learning course. The COVID-19 pandemic has provided a unique opportunity to explore this relationship, as remote learning has forced the exclusive use of ICT tools to carry out every task in the project without any other form of live interaction.

The use of ICT tools is critical in a remote learning environment. However, the proficient use of these ICT tools might also depend on the familiarity that students already have with them, as the steepness of these learning curves can trigger frustration and limit their adoption. Therefore, different students or teams can react very differently to the need to use various ICT tools during the execution of a design course structured according to the PBL pedagogy. Previous studies have limited their focus on adopting the use of tools according to the needs of different stages of the process with a cross-sectional perspective, neglecting the progression of the usage of these ICT tools during the project. In this chapter authors offer an original view on the usage of ICT tools, aiming to unveil similarities and differences among individuals and among design teams during the execution of a PBL course organised around a design project.

Understanding the relationship between individual and collaborative use of ICT tools in PBL design courses is crucial for maximising the advantages of PBL courses. Therefore, educators and ICT tools developers must comprehend how individual and team activities interact in PBL courses. By studying how ICT tools evolve throughout the design process, educators can offer targeted assistance to students, recognise possible challenges, and create a set of ICT tools that meet the requirements of design teams spread across different locations.

In the chapter **"European Global Product Realisation: Creativity and Innovation in Educating Engineers and Product Designers of 21st Century"**, Kovacevic and co-authors present a Collaborative Design in a Virtual Environment (CODEVE). This chapter provides a broader perspective on the PBL courses describing a methodology developed through years of collaborative courses simultaneously offered at different universities. Over several years, this teaching methodology originated from the European Global Product Realization (EGPR) course. With the rise of globalization, today's products are more widely distributed, and early professional practice requires students to develop global products within distributed organizations. The Global Product Realization course was created and has been performed in collaboration with European universities since the beginning of the twenty-first century. The CODEVE teaching methodology enables students to work on an industrial project, encouraging them to understand and explore methods from other disciplines while helping them overcome the barriers of a distributed environment. In such projects, students also experience communication style, relationships with teammates, and the availability and clarity of shared information that play a crucial role in the project's realization. The presented content is relevant to educators/researchers in engineering education, professionals, and developers of ICT tools for designing.

Innovation and creativity are the focal points of the third part of this book. New product development (NPD) is a complex process that involves multiple stakeholders and expectations. One of the significant issues with product development is the dynamic changes in requirements. With technologies, attitudes, and expectations of customers, management, employees, and society constantly evolving, developers must anticipate the future situation and its boundary conditions to create products that will become and remain relevant in the market. Depending on the market, time horizon, and technologies used, developers must balance developing products that meet current needs and creating products that can adapt to changing requirements.

Addressing human behaviour is another significant challenge. Individual characteristics, as well as the corporate culture, shape the behaviour of staff members in the industry. For example, continuing earnings pressure on the top management may lead to a risk-avoidance strategy, especially for capital companies. At the same time, strict control of budgets and projects can hinder staff from trying out new ideas up to a certain level. Barriers to innovation can also arise due to poor failure culture and a lack of trust between partner companies and sub-suppliers.

In the chapter **"Creativity—a Bottleneck in Engineering Design?"** Udo Lindemann discusses the innovation barriers through several research project cases. Recognising the barriers hindering innovation, such as risk avoidance at the management level, strict controlling of budgets and projects, hindering the staff from trying out new ideas up to a certain level, and making decisions based on insufficient information and knowledge, are among the major challenges for companies. Hierarchies within organizational structures often impede the exchange between specialists in different departments and are frequently one of the reasons for prolonged decision processes. Other significant barriers include missing knowledge about customers' real needs ("we know better") and a lack of competencies. Collaboration across different departments and disciplines may cause problems due to different cultures and missions. Mixing up strategy and operation, the Not-Invented-Here effect or colleagues' resentments are some additional barriers.

The challenges in product development may arise from different sources apart from the company or technology. In addition, the challenges may have different contexts like cultural, perceptual, physical or regulatory. These aspects are discussed in the **"Basics of Integrated Design Engineering (IDE)"** chapter, prepared by Sándor Vajna.

The cultural context emerges from cultural values, among these lifestyle, habits, communication patterns, or technological development of a culture. Cultural aspects affect the humans within the product life cycle, including the way of thinking, self-perception, communication patterns, beliefs, and value systems. The perceptual context refers to how people perceive and comprehend the world around them. Providers and customers with different cultural backgrounds might differ in their perceptions and attitudes towards products. The physical context refers to the geographical and psychological location where a product is developed and used. Finally, the regulatory context covers laws, regulations, standards, and guidelines that govern a product's genesis, application, and re-utilisation in its target environments on different levels of strictness. Customer demands synthesise immediate customer needs, requirements, necessities, experiences, desires, open/hidden expectations, assumptions, beliefs, peer pressure, market opportunities, social, cultural, external and internal conditions, and constraints.

IDE development attempt to meet mentioned multi-disciplinary needs, requirements and constraints. The chapter illustrates the characteristic of IDE, emphasising the focus on human centricity, which is defined as both respect and consideration of all interests, issues, needs, and matters of all humans involved with a product throughout its whole life cycle. IDE enables the generation, distribution, use, service, and re-utilisation of a product with appropriate processes that assure human centricity. The requirements for assuring human centricity must be added to the requirements resulting from the expected product performance described by the attributes.

In the chapter **"Human-Centered Design Methodology as Bridge Between Academic Research and Requirements in Industry"**, Petra Badke-Schaub and Harald Schaub present another view on the design methodology development. Human-Centred Design deals with the similar issues as Human Factors and Ergonomics, focusing on physical, cognitive, and social ergonomics. It is an approach investigating decision-making processes, the influence of technical aids and tools, information systems, information overload, and the direction of attention, among other factors. By considering the physical, cognitive, and social perspectives, Human-Centered Design can support product development and ensure that safety is integrated throughout the product's life cycle.

The role of the designer is crucial in the Human-Centered Design Methodology approach. They need to make prognoses about the user's behaviour under certain conditions and forecast the behaviour of potential users in terms of use and misuse. Established methods such as scenario thinking, systems thinking, and creativity techniques can be used. However, designers also need to have a basic understanding of psychology, especially in the field of human factors, to ensure that the design considers the cognitive and social peculiarities of the user.

By bridging the gap between academic research and industry requirements, the presented methodology offers several possibilities to derive requirements for human-system interaction, educational and training situations, and workplace optimization based on empirical research and theory from cognitive and social psychology. The quality of design process and the quality of the outcome, the product might be considered as a core concern of Human-Cantered Design Methodology.

The final contribution of this volume delves into the subject of the quality of design research. While the concept of quality finds practical and measurable criteria within the design process or the resulting design product, quality assessment in design research remains an ongoing challenge, e.g. see P.E. Vermaas discussion in "Design Science: Why, What and How" (Papalambros, 2015). Academic design research quality criteria vary across design domains, funding bodies, and academic institutions. In the chapter titled **"What Does It Mean: "Quality of Design Research"?"**, Christian Weber explores the concept of quality in design research as the degree to which a set of inherent characteristics fulfils specific requirements. This involves treating design research as a systematic inquiry that aims to discover and interpret facts, challenge existing theories or principles, and develop new applications of available knowledge to enhance the current situation. The quality of design research depends on the characteristics of research in general, such as being rigorous, empirical, and critical, as well as the specific requirements for design research, including establishing knowledge in the field, providing guidelines for scientific reasoning, and creating coherent descriptions of products and design processes. By meeting these requirements, design research can contribute to improving the current situation in design and increasing our understanding of the phenomena of design.

The closing chapter briefly summarizes the key findings and discussions from the workshop that explored various aspects of design research, education, and practice. It acknowledges the significant growth, consolidation, and diversification of design research over the years. The chapter emphasizes the multidimensional nature of design, encompassing technical, social, and policy attributes. It recognizes the immense challenges practitioners and researchers face in meeting the evolving needs and expectations of stakeholders from industry, users, policymakers, and society.

References

Albers, A., Alink, T., Thau, S., Matthiesen, S. (2008). Support of system analyses and improvement in industrial design through the contact and channel model. In: Marjanović, D. Štorga, M. Pavković, N. & Bojčetić, N. (eds.). Proceedings of the DESIGN 2008, 10th International Design Conference, University of Zagreb, Zagreb.

Alger, J. R. M., & Hays, C. V. (1964). *Creative Synthesis in Design.* Prentice Hall.

Amaresh, C. A., & Lindemann, U. (2015). Impact of design research on industrial practice tools. *Technology, and Training, Springer Cham.* https://doi.org/10.1007/978-3-319-19449-3

Andreasen M. M., & McAloone T. C. (2008). *Applications of the Theory of Technical Systems—Experiences From the "Copenhagen School,* AEDS (Workshop, Pilsen).

Andreasen, M. M., Howard, T.J., & Lomholt Brun, H. P. (2014). Domain theory, its models and concepts in A. In: Chakrabarti, L. T. M. Blessing (eds.), *An Anthology of Theoriesand Models of Design.* https://doi.org/10.1007/978-1-4471-6338-1_9 Springer-verlag, London

Andreasen, M. M. (2011). 45 Years with design methodology. *Journal of Engineering Design, 22*(5)

Archer, L. B. (1954) Artist versus engineer. *Design, 67*, 13–16.

Archer, L. B. (1963d). Systematic method for designers: Part four: Examining the evidence. *Design, 179*, 68–72.

Archer, L. B. (1963a). Systematic method for designers: Part one: Aesthetics and logic. *Design, 172*, 46–49.

Archer, L. B. (1963c). Systematic method for designers: Part three: getting the brief. *Design, 176*, 52–57.

Archer, L. B. (1963b). Systematic method for designers: Part two: Design and system. *Design, 174*, 70–74.

Archer, L. B. (1964a). Systematic method for designers: Part five: the creative leap. *Design, 181*, 50–52.

Archer, L. B. (1964b). Systematic method for designers: Part six: The donkey work. *Design, 185*, 60–63.

Archer, L. B. (1964c). Systematic method for designers: Part seven: The final steps. *Design, 188*, 56–59.

Archer, L. B. (1981). *A View of the Nature of the Design Research in Design: Science: Method.* In: R. Jacques, J. A. Powell (eds.), Guilford, Surrey: IPC Business Press Ltd.

Asimov, M. (1962). *Introduction to design.* Englewood Cliffs, New York, Prentice Hall

Bayazit, N. (2004). Investigating design: a review of forty years of design research. *Design Issues,* 20(1), pp 16-29), MIT. https://doi.org/10.1162/074793604772933739

Birkhofer, H. (2011). *The Future of Design Methodology.* Springer-Verlag London Limited. https://doi.org/10.1007/978-0-85729-615-3

Bischoff, W., & Hansen, F. (1953). *Rationelles Konstruieren Konstruktionsbücher*. Bd. 5. Berlin: VEBVerlag Technik.

Blessing, L. T. M. (2002). What is this thing called design research? *Annals of the (2002 international CIRP Design Seminar*, Hong Kong, 16–18 May, pp. 1–6.

Boyd Davis, S., & Gristwood, S. (2016). The structure of design processes: Ideal and Reality in Bruce Archer's (1968 Doctoral Thesis. In *Proceedings of DRS 2016, Design Research Society 50th Anniversary Conference.* Brighton, UK.

Brown, T. (2008). Design thinking. *Harvard Business Review, 86*, 84–92.

Cash, P. (2020). Where next for design research? *Understanding Research Impact and Theory Building, Design Studies, 68*, 113–141. https://doi.org/10.1016/j.destud.2020.03.001

Chakrabarti, A., & Blessing, L. T. M. (eds.). (2014). *An Anthology of Theories and Models of Design, Philosophy, Approaches and Empirical Explorations.* Springer. https://doi.org/10.1007/978-1-4471-6338-1

Cross, N. (2006). *Designerly Ways of knowing.* Springer-Verlag.

Cross, N. (1993). A history of design methodology. In: M. J. de Vries et aL (eds.), *Design Methcdology and Relationships with Science*, (pp. 15–27). Kluwer Academic Publishers.

Culley, S. J., Allen, R. D., & Hicks, B. J. (2002). Informal information in engineering design—a classification. In *Proceedings of DESIGN (2002, the 7th International Design Conference,* Dubrovnik.

Culley, S. J. (2014). Revisiting design as an information processing activity. In: A. Chakrabarti, L. T. M. Blessing (eds.), *An Anthology of Theories and Models of Design.* https://doi.org/10.1007/978-1-4471-6338-1_18. Springer-Verlag London.

Dixon, J. R. (1966). *Design Engineering: Invetiviness, Analysys and Decision Making*. New York McGraw-Hill.
Ehrlenspiel, K. (1985). *Kostengünstig Konstruiren*. Springer.
Erkens, A. (1928). *Beiträge Zur Konstruktionserziehung. z. VDI, 72*(1928), 17–21.
Farzaneh, H. H., Lindemann, U. (2018). *A Practical Guide to Bio-inspired Design*. Springer. https://doi.org/10.1007/978-3-662-57684-7
Feilden, G. B. R. (1963). Engineering design: report of a committee appointed by the council for scientific and industrial research to consider the present standing of mechanical engineering design. *By Council for Scientific and Industrial Research (Great Britain)*.
Finger, S., & Dixon, J. R. (1989). A review of research in mechanical engineering design. Part II: Representations, analysis, and design for the life cycle. *Research in Engineering Design, 1*, 121–137 (1989) https://doi.org/10.1007/BF01580205
Finger, S., & Dixon, J. R. (1989a). A review of research in mechanical engineering design. part I: descriptive, prescriptive, and computer-based models of design processes. *Research in Engineering Design, 1*, 51–67. https://doi.org/10.1007/BF01580003
French, M. J. (1985). *Conceptual Design for Engineers*. Springer, Berlin/Heidelberg u. Design Council, London
Friedman, K. (2003). Theory construction in design research: criteria: approaches, and methods. *Design Studies, 24*, 507e522.
Gericke, K., & Blessing, L. (2012). An analysis of design process models across disciplines. In: Marjanović, D. Štorga, M. Pavković, N. & Bojčetić, N. (eds.). Proceedings of the DESIGN 2012, 12th International Design Conference, University of Zagreb, Zagreb.
Gero, J. (1990). Design prototypes: A knowledge representation schema for design. *AI Magazine, 11*, 26–36.
Glegg, G. L. (1973). *The Science of Design*. Cambridge University Press.
Gordon, W. J. J. (1961). *Synectics: The development of creative capacity*. Collier, New York.
Hales, C. (2005). Adding value to design research. In: Clarkson, J. & Huhtala, M. (eds.), *Engineering Design Theory and Practice*. EDC University of Cambridge.
Hansen, F. (1966). *(1966) Konstruktionssystematik*. VEB Verlag Technik.
Hansen, F. (1974). *Konstruktionswissenschaft—Grundlagen und Methoden, München*. Hanser.
Hatchuel, A., & Weil, B. (2003). *A New Approach of Innovative Design: An Introduction to C-K Theory*. In: Norell, M. (ed.), International Conference on Engineering Design (ICED'03), Stockholm, Sweden.
Horváth, I. (2004). A treatise on order in engineering design research. *Research in Engineering Design, 15*, 155–181.
Howard, T. J., Culley, S. J., & Dekoninck, E. A. (2008). Describing the creative design process by the integration of engineering design and cognitive psychology literature. *Design Studies, 29*(2), 160–180. https://doi.org/10.1016/j.destud.2008.01.001
Howard, T. J., Culley, S. J., & Dekoninck, E. A. (2010). Describing the creative design process by the integration of engineering design and cognitive psychology literature. *Design Studies, 29*(2), 160–180. https://doi.org/10.1016/j.destud.2008.01.001
Hubka, V. (1974). *Theorie der Maschinesysteme*. Springer-Verlag.
Hubka, V. (1980). *Allgemeines Vogehensmodell des Konstruirens*. Fachpresse Goldach.
Hubka, V. (1982). *Principles of Engineering Design*. Butterworth Scientific Press.
Hubka, V., & Eder, E. (1987). A Scientific approach to engineering design. *Design Studies, 8*(3), 123–137.
Hubka, V., & Eder, E. (1988). *Theory of Technical Systems: A Total Concept Theory for Engineering Design*. Springer.
Hubka, V., & Eder, E. (1992). Priniciples *of Engineering Design—General Procedural Model of Engineering Design*. London: Butterworth & Co. (1982. (reprinted in WDK series by Heurista, (1988).
Hubka, V., & Eder, E. (1996). *Design Science*. Springer Verlag.

Hubka, V. (1983). *J. Dietrich zum Konstruieren Grundlegende Gedanken*, Schriftereihe WDK 9, Heurista, Zurich.
Jones, J. C., & Thornley, D. G. (Eds.). (1963). *Conference on Design Methods.* UK, Pergamon Press.
Jones, J. C. (1966). Design methods compared 1: strategies. *Design, 212*, 32–35.
Keates, S., & Clarkson, P. J. (2003). *Countering Design Exclusion: An Introduction to Inclusive Design.* Springer. https://doi.org/10.1007/978-1-4471-0013-3
Kesselring, F. (1942). Die starke Konstruktion. VDI-Z. 86 pp. 321–330, 749–752.
Kesselring, F. (1951). *Bewertung von Konstruktionen.* VDI-Verlag.
Kesselring, F. (1954b). *Technische Kompositionslehre.* Springer.
Kesselring, F. (1955). Morpfologisch-analytische Konstruktionsmethode. *VDI-Z, 97*, 327.
Kesselring, F. (1954). *Technische Kompositionslehre.* Berlin Göttingen. Heidelberg, Springer.
Koller, R. (1985). *Konstruktionslehre für den Maschinenbau.* Springer, Berlin/Heidelberg.
Krause, D., Heyden, E. (eds.). (2022). *Design Methodology for Future Products—Data Driven, Agile and Flexible.* Springer. https://doi.org/10.1007/978-3-030-78368-6
Kroes, P., Vermaas, P. E., Light, A., & Moore, S.A. (2008). Design in Engineering and architecture: towards an integrated philosophical understanding. In: *Philosophy and Design.* Springer, Dordrecht. https://doi.org/10.1007/978-1-4020-6591-0_1
Lindemann, U. (2006). *Methodische Entwicklung technischer Produkte: Methoden flexibel und situationsgerecht anwenden.* Springer.
Lindemann, U. (2009). *Structural Complexity Management*. Springer.
Lindemann, U. (1998). *Human Behaviour in Design Individuals, Teams, Tools.*
Lloyd, P., & Bohemia, E. (eds.). (2016). In: *Proceedings of DRS2016: Design + Research + Society—Future-Focused Thinking*, Vol. 1–10, DRS. https://www.drs2016.org/ddr8
Marjanović, D. (1996). Design process representation for the development of an ICAD system". *AUTOMATIKA, 37*(3–4), 127–132.
Marjanović, D., & Pavković, N. (1995). *The Structure of an ICAE System, International conference on engineering design—ICED 95*, Prague.
Matousek, R. (1957). *Konstruktionslehre des allgemeinen Maschinenbaus.* Springer, Berlin.
McMahon, C., Giess, M., & Culley, S. (2005). Information management for through life product support: the curation of digital engineering dana. *International Journal of Product Lifecycle Management 1*(1): 1743–5129 ISSN (online).
Niemann, G. (1975). Maschinenelemente, Bd. 1. Berlin: Springer 1. Aufl. (1950), 2. Aufl. (1965), 3. Aufl. (1975) (unter Mitwirkung von M. Hirt).
Olesen, J. (1992). *Concurrent Development In Manufacturing—Based Upon Dispositional Mechanisms.* Dissertation, Technical University of Denmark.
Osborne, A. F. (1963). *Applied Imagination—Principles and Procedure of Creative Thinking.* Scribners, New York.
Pahl, G., & Beitz, W. (1977). *Konstruktionslehre.* Springer-Verlag.
Pahl, G., Beitz, W., Feldhausen, J., & Grote, K. H. (2007). *Engineering Design: A Systematic Approach.* Springer.
Pahl, G., & Beitz, W. (1984). *Engineering Design.* In: Wallace, K.M. (ed.)(1st edn), The Design Council, London.
Papalambros, P. (2015). Design science: why, what and how. *Design Science, 1*, E1. https://doi.org/10.1017/dsj.2015.1
Petric Maretic, H. (2014). Dynamic Network Analysis—*Evolution of the Research Field Based on the Papers Published at the International DESIGN Conference 2002–2014 (in Croatian)* URL: https://cadlab.fsb.hr/en/education/projects/dynamic-network-analysis-evolution-of-the-research-field-based-on-the-papers-published-at-the-international-design-conference-2002-2014-5
Ponn, J., & Lindeman, U. (2011). *Konzeptentwicklung und Gestaltung technischer Produkte Systematisch von Anforderungen zu Konzepten und Gestaltlösungen.* Springer.
Reuleaux, F. (1861). *Der Constructeur*. Vieweg.

Reuleaux, F. (1875). *Theoretiche Kinematik, Braunschweig, Vieweg,* (1875/1900).
Reuleaux, F. (1891). *Technology and Civilization, From the Smithsonian report for 1890.* Washington, Gov. Printing office.
Rodenacker, W.G.; Claussen, U. (1973). *Regeln des Methodischen Konstruierens/T*. 1. Regeln 1 bis 4. Mainz: Krausskopf.
Rodenacker, W.G., Claussen, U. (1975). *Regeln des Methodischen Konstruierens/Regeln 5 bis 8.* Mainz: Krausskopf.
Rodenacker, W.G. (1970). *Methodisches Konstruieren. Konstruktionsbücher*, Bd. 27. Berlin: Springer.
Roozenburg, N. F. M., Eekels, J. (1995). *Product Design: Fundamentals and Methods.* Wiley, Chichester.
Roth, K. (1982). *Konstruieren mit Konstruktionskatalogen.* Springer, Berlin/Heidelberg.
Simon, H.A. (1969). *The Sciences of the Artificial*. The MIT Presss.
Suh, N.P. (2001). *Axiomatic Design.* Oxford University Press.
Tjalve, E. (1976). *A short course in industrial design.* Newnes-Butterworths.
Tomiyama, T. & Yoshikawa, H. (1986). Extended general design theory. Technical Report CS-R8604, Centre For Mathematics and Computer Science, Amsterdam.
Ulrich, K.T., & Eppinger, S.D. (2004). *Product design and development*. McGraw-Hill/Irwin.
Wallace, K., & Blessing, L. (2000). Observations on some German contributions to engineering design in memory of professor Wolfgang Beitz. *Research in Engineering Design, 12*, 2–7. https://doi.org/10.1007/s001630050019
Weber, C., Birkhofer, H. (2007). Today's Requirements on engineering design science. In: *Proceedings of ICED 07:785–786* (*Exec. Summary*), Paper-no. 511 (Full Paper, CD-ROM), École Centrale Paris.
Weber, C. (2005). CPM/PDD—an extended theoretical approach to modelling products and product development processes. In*: Proceedings of the 2nd German-Israeli Symposium on Advances in Methods and Systems for Development of Products and Processes*, pp. 159–179. Berlin, Germany, Fraunhofer-IRB-Verlag, Stuttgart.
Weber, C. (2008). How to derive application-specific design methodologies. In: *Proceedings of DESIGN*, vol. 1, pp. 69–80, Faculty of Mechanical Engineering and Naval Architecture, University of Zagreb
Wögerbauer, H. (1943). *Die Technik des Konstruierens* (2nd ed.). Oldenbourg.
Yoshikawa, T. (1981). General Design Theory and a CAD system. In: Sata, T. and Warman, E. (eds). Man-Machine Communication in CAD/CAM, Proceedings of The IFIP WG5.2–5.3 Working Conference (1980), pp. 35–37. North-Holland, Amsterdam.
Zwicky, F. (1966). *Entdecken.* Forschen im Morphologischen Weltbild, München/Zürich, Droemer Knaur Verlag.

Chapter 2
From Design Optimization to Design Science: An Evolution in Design Thinking

Panos Y. Papalambros

Abstract Design optimization evolved in parallel to operations research as a way to codify and support design decisions mathematically. Design thinking emerged as a way to describe a user-centered design process that seeks to unpack the core values behind design decisions. In the modern definition of Design Science—as the field that studies the creation of artifacts and their embedding in our physical, psychological, economic, social, and digital environments—the two original approaches are merging. We offer a perspective on the design optimization evolution that links the engineering, business, computer, behavioural, and public policy sciences, primarily through mathematical modelling while explicitly recognizing its limitations.

2.1 Introduction

Design as a decision-making process has been a popular and powerful paradigm for engineering design (see, e.g., Siddall, 1972; Lewis et al., 2006). The power of this paradigm lies on the importance of a cognitive process that is core to design, namely, making choices among alternatives. All stakeholders, from designers to users, make choices throughout the design process and use stages. Adopting the decision-making paradigm has allowed design researchers and practitioners to access and use all of the developments, theories and techniques in decision theory, decision analysis, operations research, game theory, and optimization. In particular, it has enabled the use of mathematical formulations of all kinds in a design context, often referred to as design automation. Design optimization has been recognized as a major area of research and practice within design automation in the past four decades (see, e.g., Papalambros & Wilde, 2017).

Choices require the presence of alternatives. Creating design alternatives is the process attendant to decision making. This creation process is less amenable to mathematical modeling and more directly linked to human cognitive processes such

P. Y. Papalambros (✉)
Department of Mechanical Engineering, G.G. Brown Laboratory, University of Michigan, Hayward, Ann Arbor, MI 48109, USA
e-mail: pyp@umich.edu

D. Marjanović et al. (eds.), *Design Research: The Sociotechnical Aspects of Quality, Creativity, and Innovation*, https://doi.org/10.1007/978-3-031-50488-4_2

as creativity. The relevant cognitive science research fits more in the domain of psychology than engineering, though both fields have clearly a stake in it. Computer science has also an increasing role in the creation of alternatives. Design thinking emerged as a formalism of successful practice in identifying the right design problem, creating alternative solutions, and making choices largely through rapid prototyping (see, e.g., Brown, 2009; Cross, 2011). Dealing with cognitive processes—being thinking after all—avoided analytical-quantitative approaches in the original development and practice of design thinking.

Early on engineers were close to the origins of the needs that defined the problem and therefore the scope of the engineering task, but gradually the increased complexity of engineering design tasks directed engineering designers to focus on the machines that were to be the solution to whatever problem was presented to them. Industrial (product) designers were relatively free from the complexity of engineering, including analysis, and focused more on the impact of a created artifact on the user experience with respect to user needs and desires. Design thinking, as a formalism of user-centered design practice, emerged in stark contrast to engineering design obsessed with functionality. A tension between solving the right problem and solving the problem right is still present in design organizations. Even in architecture where the coupling between functionality and aesthetics has been well understood for a long time such tension is not absent.

Decisions have outcomes. The outcome of a designer's decisions is placing an artifact in the world for possible use, where "artifact" here means a product, system or service, or a combination of them. The outcome of a user's decision is to select this artifact and indeed use it in the world. This use has consequences the designer or user may have not intended or anticipated, particularly when such use becomes large scale and for extended periods of time. What might have been the needs and desires that the original artifact intended to satisfy have now become realities that everyone must live with—individual humans, groups of humans, societies, and diverse species in the physical world—whether they made the choice or not. We often refer to these outcomes as unintended consequences. This term tends to be self-serving for both creators and users as it implies an intellectual hand washing of the responsibility for *unintended outcomes* (see, e.g., Gelles, 2018). The well-established tragedy of the commons phenomenon in environmental sustainability reflects exactly the avoidance of individual responsibility. Early scientific pursuit of artifact creation as articulated from a computer science perspective (see, e.g., Simon, 1988) was rather silent on the consequences of *embedding* the artifacts in the world.

Who then has the responsibility for the impact that artifacts have in the world we live—in the *designed world*?

The obvious answer is that everyone involved in the decision making that creates the designed world is responsible. Everyone is a designer! This means that all participants in the creation of the designed world must do their part to make it the best we can. Inevitably then, science as the highest human intellectual achievement must be brought fully to bear on designing the designed world. Thus follows the modern definition of design science as the discipline that studies the creation of artifacts and their embedding in our natural, virtual, psychological, economic, and social environments. While design is often linked with artifact creation (assiduously often taken as synonymous to innovation), artifact embedding is critical for the modern study and practice of design.

In this chapter we trace the evolution of design optimization from its birth in engineering design to its gradual maturity within the broader design science over the past four decades. The evolution described here closely follows the author's experience and is not meant to be a literature review; rather, it aims to serve as testimony on the progress of this evolution to date and as incentive to sorely needed future design research, thinking and practice. In the remainder, we examine briefly the design optimization construct and then we illustrate how this construct has been used to integrate behavioural, social and policy considerations, thus demonstrating how science from different domains can be integrated into design science. We conclude with discussion on limitations and challenges.

2.2 Design Optimization

Design optimization is an engineering design methodology that uses a mathematical formulation of a design problem to support selection of the optimal design among many alternatives. The typical (so-called negative null) formulation is stated as (Papalambros & Wilde, 2017),

$$\begin{aligned} &\text{minimize} && f(x) \\ &\text{subject to :} && h_i(x) = 0, \quad i = 1, 2, \dots, m_1 \\ & && g_j(x) \leq 0, \quad j = 1, 2, \dots, m_2 \\ &\text{and} && x \in X \subseteq R, \end{aligned} \tag{2.1}$$

where $\mathbf{x} = (x_1, x_2, \ldots, x_n)$ is a vector of n design variables with values in the real number set R,$f(\mathbf{x})$ is the objective function, $h_i(\mathbf{x})$ are m_1 equality constraints, $g_j(\mathbf{x})$ are m_2 inequality constraints, and X is a set constraint that includes additional restrictions on $\mathbf{x}$ besides those expressed by the equality and inequality constraints.

The first challenge is the translation of a design problem into this form. This translation requires two key steps:

1. Describing alternative designs through a set of design variables, so that assigning different values to these variables generates different designs alternatives.
2. Determining the mathematical functions for the design objective and constraints, so that these functions can be computed for different values of the design variables.

As we will discuss further below, these steps require both creativity and skill in analysis of the phenomena involved in the particular design problem. Moreover, the optimal solution computed is only optimal within the confines of the model, and for many reasons there can be design considerations not included in the model.

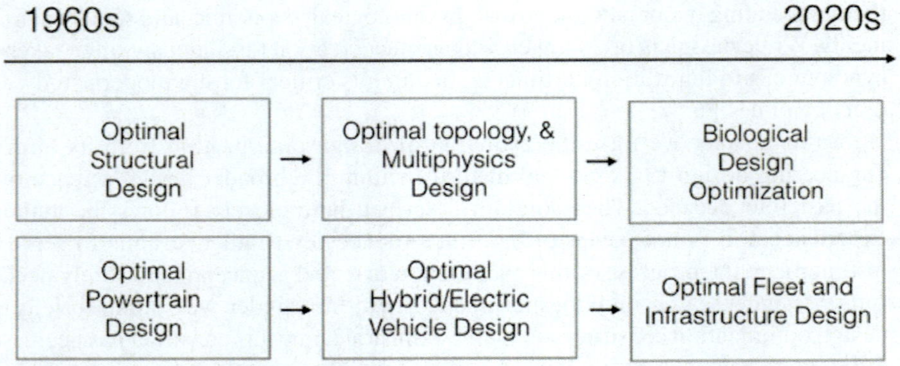

Fig. 2.1 Evolution examples in engineering design optimization

The rapid progress and commercialization of tools in computer-aided engineering, such as computational mechanics, have enabled us to predict the functionality of alternative designs (Step 2 above) with high accuracy.

The second challenge is solving correctly the resulting mathematical problem. In the formulation above we made several assumptions already. For example, we assumed that the design decisions can be represented by a vector of real-valued variables; however, there may be problems where $\mathbf{x}(s)$ with s being a state variable such as time in controls problems or spatial location in variational calculus. Strategies, algorithms, and computation implementations for solving optimization problems occupied the best part of four decades, roughly 1960–2000, in a broad range of disciplines including design optimization. Algorithmic refinements, along with computation power and commercial implementations, have successfully addressed this second challenge for many practical problems.

Within engineering design optimization, the evolution followed our increased ability in both modeling and computation. We will look next at two examples, structural and powertrain design, see Fig. 2.1.

2.2.1 *Structural Design*

In the domain of structural design, early formulations followed variational calculus models and were limited to problems were the resulting optimality criteria (i.e., stationarity conditions) were solvable analytically or with relatively simple computation. Early formulations of mathematical programming models, such as linear and convex programming, for structural truss problems opened up the range of problems that could be addressed. Optimality criteria methods also became more versatile with more sophisticated computational methods using discretization techniques. For a while, structural optimization was dominated by the debate between what is best: mathematical programming or optimality criteria. As the field matured the debate subsided with the recognition of the essential equivalence of the two approaches (Fleury, 1982), one performing discretization when posing the problem (mathematical programming), the other when solving it (optimality criteria).

The above debate was particularly influential in the evolution of problem sophistication. Early structural design problems were usually proportionality or sizing problems, namely, fixed shape and topology but with variable sizes like lengths, thicknesses and so on. Variational formulations allowed for shape optimization while mathematical programming formulations struggled to materialize efficient solutions. Both failed to address successfully the topology design problem. For example, to optimize structural truss design we could deal with topology by deleting truss elements but not by adding. So the strategy, eventually called the ground structure approach, was to fix the nodes of the truss and assume all possible links exist, then gradually remove the unnecessary links. In the next step we were able to also move the node locations.

For continuous structures though, the struggle for topology optimization continued until a profoundly simple and powerful idea emerged: Instead of looking for sizes and locations as the design variables, we could use instead the distribution of material in the designated design space and make the material density (e.g., from zero to one) in each voxel in space as *the* variables (Bendsøe & Kikuchi, 1988). The resulting formulation of minimizing structural compliance subject to a global constraint on available material could then be solved through discretization by a standard mathematical programming algorithm like sequential linear programming (Bendsøe, 1995).

This simple but powerful idea opened up the topology design optimization problem to consider other non-structural problems, since all you needed was the existence of suitable tensor field equations besides elasticity, such as electromagnetic field. Moreover, what you could do in the macro scale, you could also do in the micro scale of material design. Design optimization now included the so-called multi-physics problems where field equations from different disciplines could be combined to design new materials and structures (see, e.g., Dede et al., 2014).

Around the same time, we had the emergence of stereolithography and layered manufacturing. For the first time we were able to build complex structures computed through topology design optimization that included voids such as porous bones. The evolution of what we now call additive manufacturing or 3D printing that enables building not just porous structures but multi-material structures has dovetailed well with the evolution of design optimization. Indeed, the design-manufacturing integration that had been pursued for several decades has taken a new meaning in that we can now look for simultaneously optimizing the design of material, structure and fabrication process (e.g., Chen et al., 2017). This ability has become most attractive in biological applications (e.g., Hollister et al., 2002; Murphy & Atala, 2014) which will remain a research frontier for the foreseeable future.

The key point here is that the above topology design capabilities broached analytically what had been thus far a problem whose solution was based solely on human creativity and ideation. Human involvement is still needed not just for the right model formulations but also for accounting for the many design concerns not included in the models.

2.2.2 *Powertrain Design*

Automotive Powertrains (PTs) presented an attractive domain for design optimization. Design of Internal Combustion Engines (ICE) became a focus of high scrutiny following the oil crisis of the 1970s. A clear emphasis on increasing fuel economy had to be balanced with regulations on emissions and a market push for performance such as acceleration. Significant gains in fuel economy could be realized with "lightweighting," namely more efficient structural design and material selection to reduce weight while meeting various safety regulations like crashworthiness and rollover protection (e.g., Luk et al., 2017). In this respect, the above evolution in structural optimization played a major role and has been incorporated in most modern commercial software.

The other path for improving fuel economy while meeting emissions and performance metrics required further ICE refinement, for example, optimizing valve timing, exhaust manifolds, cylinder heads, cam profiles and other engine component geometry to increase overall engine efficiency while accounting for the aforementioned constraints (see, e.g., Wagner and Papalambros, 1999). Such design optimization was largely enabled by the development of computational fluid dynamics and other computational mechanics tools that could be used to predict design performance and hence identify optimal solutions. Experimental validation and calibration was then much simplified in both time and cost.

PT control or "energy management" strategies were important in ICE-based PT design but they became critical with the advent of hybrid powertrains. A powertrain that combined an ICE with electric motor-generators, particularly in parallel hybrid modes, required deeper consideration of the control strategies needed to switch power delivery in real time. The PT design problem became a coupled optimal design—optimal control problem and it was no longer acceptable to solve the two problems separately. A poor design could make a good control strategy impossible. Thus PT optimization became a combined control and design, or *co-design*, problem (Alyaqout, 2006; Fathy, 2003; Peters, 2010; Reyer, 2000). Co-design problems had been studied relatively little and in specific domains, for example, in deployable tensegrity structures for space applications. Evidently, co-design optimization would be required in the development of all so-called smart products and systems, for example, in robotics. Early formulations of co-design problems used nonlinear programming models for proportionality design optimization coupled with gains as the control variables for fixed controller topologies. Generalization to include optimal control strategies still assumed a known controller topology.

The success enjoyed in structural and multi-physics topology design did not spill over to other design domains, notably electromechanical systems design. The key impediment has been our inability to capture the design space of system topologies under a single model, as we had done in structural topology. Therefore, designer intuition and human creativity remained the primary source of design ideation for such systems. However, in a specific domain such as Electric Vehicle (EV) and Hybrid-Electric Vehicle (HEV) powertrain design limited success was achieved in formulating topology co-design models. This was accomplished through a modification of bond graphs that allow modelling EV/HEV topologies under a single model (Bayrak, 2015; Bayrak et al., 2016). Optimizing EV/HEV PT configurations (another name for 'topologies') is important because different topologies are appropriate for different vehicle types. Having a formal exploration of PT configurations is a significant aid to PT designers, similarly to structure designers. Extensions to broader elecromechanical topology design problems are still lacking.

Armed with the above EV and HEV PT system design capability, the next evolutionary step was to link the design of these new vehicles with the design of larger systems, such as microgrids and vehicle fleets (Ersal et al., 2013). Electric energy microgrids are considered for civilian applications such as isolated communities without a central grid or isolated military and emergency operations. Such microgrids can interact with EVs as sources of both power and transportation, depending on the needs of the moment. Vehicle fleet design is part of infrastructure, urban or suburban, design as well as of military and emergency mission planning (Bayrak et al., 2018a). The idea is that we can look at the needs (or set of missions) of the particular community and determine the vehicle type mix and associated vehicle designs as a large-scale design optimization problem (Bayrak et al., 2018b). This 'system of systems' design is now executed at a high level of performance fidelity where relatively detailed individual vehicle design is executed together with the fleet design to optimize an overall community objective such as cost, energy use, or convenience. The current studies of autonomous vehicles and fleets will only increase the value of such integrated system design optimization.

The above two evolutionary paths of design optimization are summarized in Fig. 2.1.

2.2.3 *Evaluations, Preference, and Choice*

The objective function in Eq. (2.1) implies that we have a metric to compare two different designs so that we can pick the 'better' one. This metric and therefore the definition of what is a better design drives design decisions, while satisfying the constraints in Eq. (2.1) ascertains that the design is feasible, namely it functions as intended. The metric itself can be an objective one in the sense that reasonable people will agree on the measurement of a particular value, such as the weight of an object. We then say that calculation of the objective function is *an evaluation.* An evaluation can be right or wrong. However, many design decisions are subjective in the sense that reasonable people may give the same design metric different values. We then say that the objective function expresses *a preference.* A preference cannot be right or wrong per se. Whether evaluation or preference, the outcome of using the metric

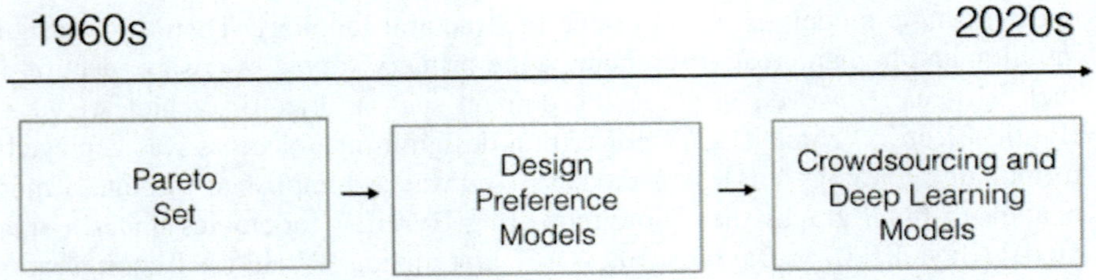

Fig. 2.2 Evolution of choice modeling in design optimization

is a choice. We will now trace how choice modeling evolved in the context of design optimization, see Fig. 2.2.

Most if not all design problems have inherently not one but many objectives. We are arguably greedy and we want everything, such as both high performance and low cost. Facing multiple competing objectives has been a well-known problem defined elegantly by Vilfredo Pareto, an engineer turned economist, in the late nineteenth century. The solution is a set of Pareto points characterized by the property that you cannot move away from them to improve one objective without making at least another objective worse. In a restatement of Eq. (2.1), the multi-objective problem is defined as:

$$\begin{aligned} &\text{minimize} && \mathbf{f}(\mathbf{x}). \\ &\text{subject to :} && h_i(\mathbf{x}) = 0, \quad i = 1, 2, \ldots, m_1 \\ & && g_j(\mathbf{x}) \leq 0, \quad j = 1, 2, \ldots, m_2 \\ &\text{and} && \mathbf{x} \in X \subseteq R, \end{aligned} \tag{2.2}$$

where $\mathbf{f}(\mathbf{x}) = (f_1, f_2, \ldots, f_m)$ is a vector of m competing objective functions.

Much work in early design optimization focused on efficient ways to generate the Pareto set. However, once the set was generated we still had the problem of selecting a single Pareto point for the final design. Most academic papers would dispense with this step by stating that the designer will make the choice using her intuition. Calculation methods, for example through some scalar substitute objective function that weighted the various objectives, could yield a single point by unwittingly hiding the essential subjective choice of weights.

Eventually, the idea of *preference elicitation* emerged: study the stakeholder making the choice to account, and hopefully understand, his preferences. An early way to do that was using an interactive genetic algorithm (IGA) where the fitness function was replaced by direct human input in a real-time interaction with a human subject (Kelly et al., 2010). Success would depend on our ability to generate new design alternatives quickly following each subject response. Moreover, the design choice would fit one particular subject but our interest would be usually in choices of a large number of subjects in a targeted population. A more general model of preference would be needed, built using choice data from this large number of subjects.

Other fields had already been at work on this problem, for example in mathematical psychology and marketing with models such as PREFMAP and Conjoint Analysis (see, e.g., Kelly et al., 2011). These models are built using a number of subjects in a controlled environment (e.g., experiments or surveys) or collecting sales data that led to 'revealed' preferences expressed through actual choices of where we spend our money, as opposed to 'stated' preference where people just say what they would choose. The design community was quick to adopt such modeling practices while recognizing some of their limitations. Design preference models became a fundamental element in design for market systems, as discussed in Chen et al. (2012) and in Sect. 2.3.

Starting with individual preference elicitation (e.g., Ren & Papalambros, 2011), the obvious desire to extend elicitation from a single subject to a large number of them, and thus form more general models, was greatly facilitated with easy access to subjects through online tools such as Amazon's Mechanical Turk (see, e.g., Ren & Papalambros, 2012). This gave rise to crowdsourcing for design, where preferences among particular tradeoffs, usually design attributes , were elicited from large numbers of subjects via some reward system. Crowdsourcing with rewards has its drawbacks in terms of the quality of derived data but also can fail for even simple design problems (Burnap et al., 2015). An alternative approach was gamification, where players would generate choice data through playing the game (Ren et al., 2016). More recently, vast amounts of such data are increasingly collected through smart devices and software applications in the 'Internet of Things'. The data thus collected can be aggregated and used to develop models though machine learning. Early machine learning models merged with modeling strategies known in mathematical psychology and statistics into what are now popularly known as deep learning models.

Simplifying somewhat, deep learning models are sophisticated 'curve-fitting' representations: Functional representations of some property, for example, visual preference, in terms of design attributes of this property, for example, shapes or colors, that *fit* a large amount of collected data. This *fitting* is itself effected through a mathematical optimization formalism. The large scale of the data and the complexity of models, such as nonlinearities, most often prevent us from securing mathematical optimality and the fit achieved is essentially a high quality heuristic. This is particularly true for even simple design problems as opposed to widely celebrated problems in computer science such as face recognition.

Another challenge for deep learning models in design is the interpretation of the derived model. In particular, the most successful deep learning methods do not assume a priory knowledge of the attributes associated with the property being modeled. Instead they use 'features' what are mathematical artifacts generated from the data (not to be confused with features in, say, feature-based design). Thus, while a functional relationship (model) is created between a property and the data used, its meaning and its predictive ability is very limited. In the current state of the art, the most value of such models is to inform qualitative insights of designers when making the relevant design decisions (see, e.g., Burnap et al. 2016).

Nevertheless, such models can be included in principle within a standard design optimization framework to represent *perceptual* (subjective) attributes side by side with the now classical engineering functionality (objective) attributes. This inclusion is critical in the evolution of design optimization to serve design science, as we will discuss in the next section. Suffice it to note that all models are approximations of reality, so the question is now more one of fidelity rather than possibility.

At this point, it is important to make clear the distinction between what the user "experiences" in an artifact and what the designer can quantify and control. *Design attributes* are the design properties that the people who experience the artifact will use to judge the artifact, e.g., easy to use, inexpensive, attractive, or well functioning. *Design characteristics* are the design properties that the designer can explicitly act upon by manipulating the design; characteristics must be measurable. For the designer to affect the user, there must be a mapping between attributes and characteristics. For example, this *design mapping* is part of the qualitative 'house of quality' in Quality-Function-Deployment.

In design optimization, the design mapping must be an actual functional mapping and the design characteristics, further classified as objectives and constrains, must be themselves functions of the design variables: $\mathbf{a} = \mathbf{a}(\mathbf{c}(\mathbf{x}))$ where $\mathbf{a} = (a_1, a_{2,} \ldots, a_n)$ is the vector of attributes, $\mathbf{c} = (c_1, c_{2,} \ldots, c_n)$ is the vector of characteristics, and $\mathbf{x} = (x_1, x_{2,} \ldots, x_n)$ is the vector of design variables. Examples of $\mathbf{a}- > \mathbf{c}$ mappings are 'ease of use' (attribute) to 'force required to move a lever' (characteristic) or 'beautiful' (attribute) to 'golden ratio proportionality' (characteristic).

Much of the design optimization evolution to design science is due to the ascendance of these design mappings, including those that are being derived from machine learning. The mapping can be simple when an attribute is directly measurable and therefore also a characteristic, or complicated when perceptual attributes are involved.

2.3 Design for Market Systems

Products do not exist without markets: someone has to commit resources to gain the value contained in the product, system or service. A first step to place design optimization within a social context was to place design decisions into a market framework. This step required shifting the design decision maker from being a 'designer' to being a 'producer,' the entity who designs, builds, and sells the product (Georgiopoulos et al., 2005, Michalek et al., 2005, Michalek et al., 2010). This shift in perspective became known as Design for Market Systems and the associated design optimization framework is shown in Fig. 2.3.

The producer's objective function is to maximize profit, where

$$\{\text{profit}\} = \{\text{revenue}\} - \{\text{cost}\} = \{\text{price}\} \times \{\text{demand}\} - \{\text{cost}\} \tag{2.3}$$

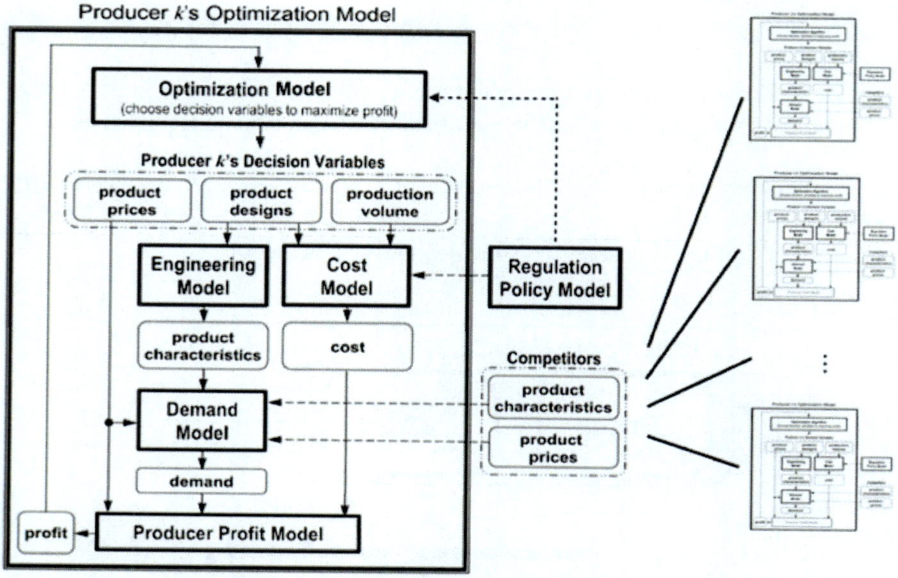

Fig. 2.3 Design for market systems framework (after Michalek et al., 2004)

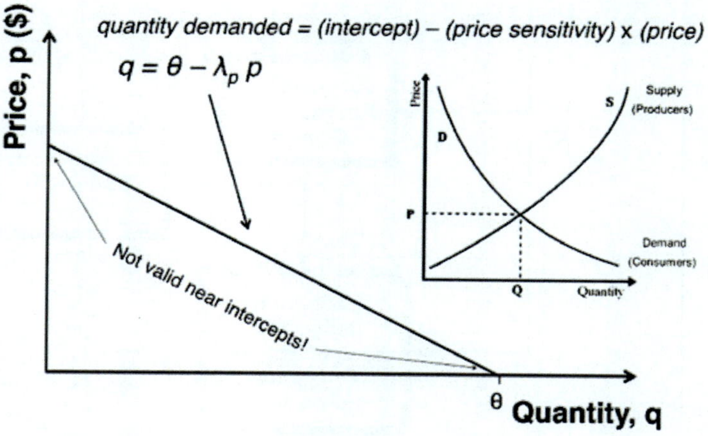

Fig. 2.4 Classic linear (inverse) microeconomic model of demand

Here price is an independent optimization variable, demand is a function of price and of the design attributes that enter into the customer's choices, and cost is a function of the same design attributes and maybe others. The classic microeconomic linear model of demand as a function of only price is $q = \theta - \lambda_P\, p$ where q *is* demand, θ is a parameter (the intercept shown in Fig. 2.4), λ_P is the price elasticity of demand, and p is price. This model had to be extended to include design attributes. A simple extension was (Papalambros & Georgiopoulos, 2006):

$$q = \theta - \lambda_P p + \lambda_\alpha \alpha \quad (2.4)$$

where $\boldsymbol{\lambda}_\alpha$ is the vector of elasticities for the $\boldsymbol{\alpha}$ vector of design attributes. Under the assumption that a producer sells what it makes, we can rewrite Eq. (2.3) as.

$$\pi = pq - c = p\,(\theta - \lambda_P p + \lambda_\alpha \alpha) - c = p(\theta - \lambda_P p + \lambda_\alpha \alpha(\mathbf{x})) - c(\mathbf{x}) \quad (2.5)$$

where π is profit, c is cost and $\mathbf{x}$ is the design variable vector.

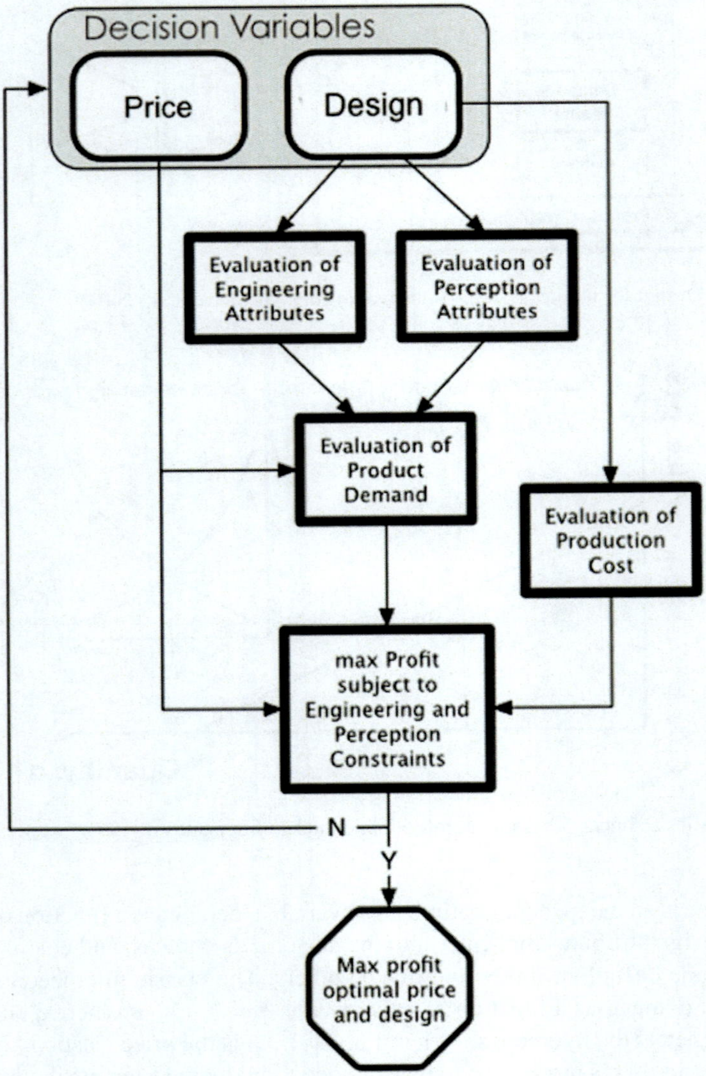

Fig. 2.5 Incorporating perception attributes along with functionality attributes

Preference models developed in marketing such as conjoint analysis were introduced as more sophisticated models of demand to replace this simple linear model. Moreover, while classical linear models of demand were derived typically from revealed preference data from sales, conjoint type of models could be built also using stated preference data from surveys and thus, in principle, model choices for designs that do not yet exist in the market.

The left part of Fig. 2.3 illustrates the connections among the different elements in one producer's design optimization model. The producer's decisions involve product price, design, and production volume. The profit model requires as inputs the values for price, demand, and cost. The demand model requires as input the values of the design characteristics and price. The optimization model maximizes profit subject to functionality constraints, typically coming from engineering requirements.

The right part of Fig. 2.3 illustrates the fact that a single producer operates within the market and under government regulations. Regulations can enter the cost model directly but also the optimization model as additional constraints. Competitors come with their own values for price and characteristics (hence, the attributes associated with these characteristics). Each competitor executes an optimization process like that of the *k* producer on the left. Taking a game-theoretic approach all producers optimize their design and price until they reach market equilibrium. Typically, Nash equilibrium has been used as the simplest to execute computationally, while recognizing the many assumptions behind it (Michalek et al., 2004).

Incorporating perceptual attributes alongside functionality attributes within the model of design for market systems is illustrated in Fig. 2.5. Here is where the choice and preference models mentioned in the earlier section become significant in predicting demand. Emotions are major contributors to choice, and so in this sense the classical design optimization model can now directly include emotional design as a contributor to decision making (Reid et al., 2010, 2012). This was a major step in the evolution of design optimization towards design science.

2.4 Public Policy

A market perspective of design decisions is an example of studying how implementing these decisions affects the social environment. As Fig. 2.3 illustrates, government policy directly affects these decisions. Policy is expressed through regulations, investments, incentives and penalties. Governments use policy to promote social good. Producers respond with products designed and priced to maximize their profit. Society responds through the political process and purchasing decisions. The automotive sector and the design of ground vehicles is a prime example of this complex interplay. Automobiles are unique products in that they are the highest-priced large-scale consumer products embodying emotional appeal while also having high impact on lifestyle, urban-suburban development, environmental degradation, and the economy at large. These attributes of the automobile design problem have motivated automotive application studies in design optimization. In this section we review how a design science perspective evolved in these studies through the incorporation of government policy.

An early example was an enterprise-wide decision problem for a publicly held firm, which seeks to determine asset allocation and asset properties concurrently as investments that maximize shareholder value (Georgiopoulos et al., 2005):

Maximize	net present value.
with respect to	asset properties (what to produce).
	asset allocation (how much to produce).
subject to	engineering constraints.
	investment constraints.

The particular study is shown in Fig. 2.6. The key regulation at the time (early 2000s) in the US market was the Corporate Average Fuel Economy (CAFE) standard that, along with market preferences, dominated products and product mix for automotive producers.

Using models similar to those discussed above, the study concluded that a producer with relatively small production capacity should strive for high fuel efficiency compact cars and average performance of the sport utility vehicles, while at relatively high capacity the number of compact cars would be high enough to mitigate the regulatory penalty and allow sport utility vehicles with 1/3 increased engine size. The CAFE regulation was an active constraint to consumer preferences, as long as consumers asked for more horsepower and producers materialized this preference in terms of profit. An important policy question raised then was whether new technologies could make the CAFE constraint inactive and at what cost.

Firm	Automotive Producer	
Resources and Capabilities	Compact Vehicle	Sport-Utility Vehicle
Opportunities	Low Profit Margin	High Profit Margin
Regulation	Build and transfer credit to higher segments	Penalty due to regulatory restrictions
Decisions	How should the firm design its products and allocate manufacturing resources	

Fig. 2.6 Automotive producer's decisions on asset properties and allocation

Another study in the early 2000s (Michalek et al., 2004) brought emissions regulations into consideration along with fuel economy, using the model in Fig. 2.3 to examine different policy scenarios for the compact car market, including CAFE, carbon dioxide emissions taxes, and diesel technology quotas. The reported results showed, for example, that CAFE standards would achieve higher average fuel efficiency per regulatory dollar, while the impact of CO2 taxes on producers for expected life cycle emissions would diminish returns on fuel efficiency improvement per regulatory dollar as the taxes increase.

At the time, there was vigorous debate and political posturing on both the government side and the automotive producers. The design for market systems framework gave the ability to compare regulations and achieve realistic trends providing insights for how varied regulations impact industry, consumers, and the environment.

In the subsequent two decades the automotive market continued to change, along with technology, policy and preferences. A recent study (Kang et al., 2016) examined the electric vehicle (EV) adoption problem: to address range anxiety (how far can the vehicle drive without refueling) EV adoption depends on vehicle price, design, and market size, as well as recharging length, cost and station availability. The interests of the three stakeholders, government, EV producer, and charging station operator, must be combined in order to quantify the effect of government investment on the EV market, see Fig. 2.7.

The study collected preference data from two markets (Ann Arbor, Michigan, USA and Beijing, China) and computed optimal public investment strategies to minimize emissions. The study found that collaboration among stakeholders can achieve both emission reduction and profitability, while EV subsidies alone would be ineffective. Also, when vehicle and station designs improve beyond a certain threshold, government investment influence on adoption is attenuated, in part because customers were

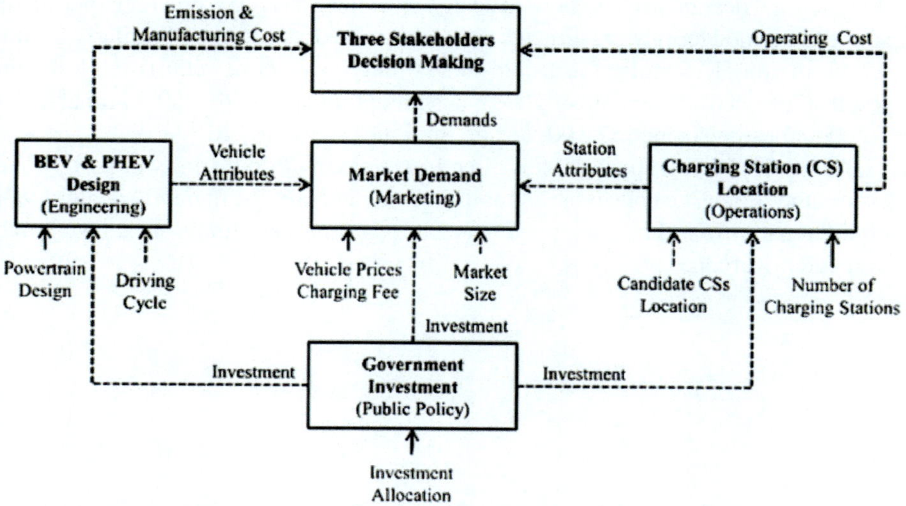

Fig. 2.7 Decision-making framework for the electric vehicle market (after Kang et al. 2016)

found to be very sensitive to price. Also, a diversified government investment portfolio would be more effective in China, with charging costs and price cuts on license plate fees being as important as EV subsidies.

The studies above use a design optimization formalism at their core. Linking design decisions from the diverse perspectives of engineering technology, marketing and finance, and government investment and policy to quantify the impact of these decisions on the world has led to the broader design science thinking.

2.5 Conclusion

Design Optimization emerged from operations research and mathematical optimization, including variational calculus. Design Thinking emerged from creativity and problem solving strategies. Design Science emerged from the designers' angst for philosophical differentiation, from the preoccupation for the meaning of 'artificial' in computer science, and from the realization that creating and living in the artificial world is a whole-science endeavor that defies single ownership.

The evolution of the design optimization formalism into design science is summarized in Fig. 2.8. Objective and subjective design attributes corresponding to functionality and perceptual attributes, have been integrated, as have functionality and human evaluation constraints. The objectives driving design decisions have evolved from purely functional ones to a producer's utility to a social value utility. This design optimization formalism with a social value utility subject to general functionality and social evaluation constraints is well-placed within design science, serving as an effective quantification approach. It is instructive to look at the evolution of the relevant terms as used in books available in Google Books shown in Fig. 2.9. Note that 'optimal design' is used for both design optimization and design for experiments, so the combined term overestimates the occurrence of the former. Similarly, a quick search in Google Scholar for articles containing these terms returns the following results (accessed 25 September 2018): design optimization 561,000; optimal design 765,000; design science 83,100; design thinking 54,500 results.

The limits of this approach are directly coupled with our limits in pretty much every other scientific endeavor: our ability to model reality sufficiently well for valid predictions. Insofar that design is a 'wicked' problem these limits are unlikely to be ever fully overcome.

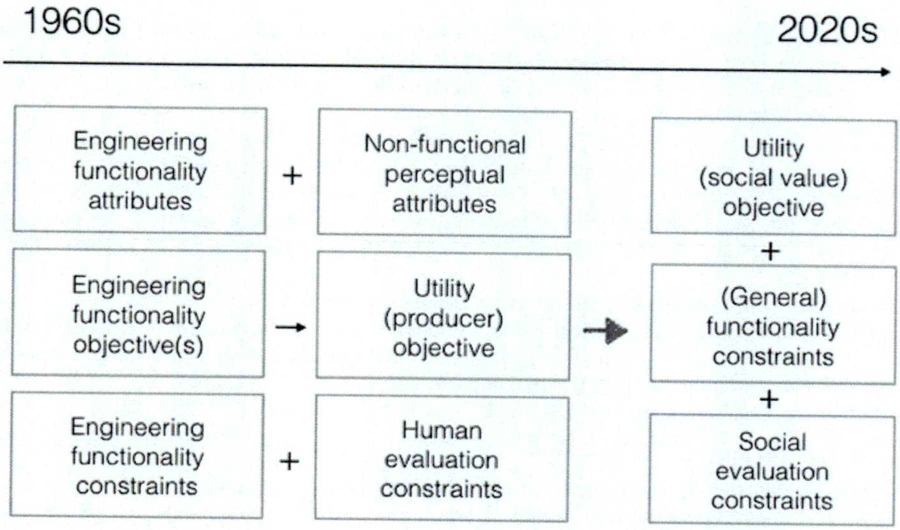

Fig. 2.8 Evolution of design thinking from engineering design optimization to design science

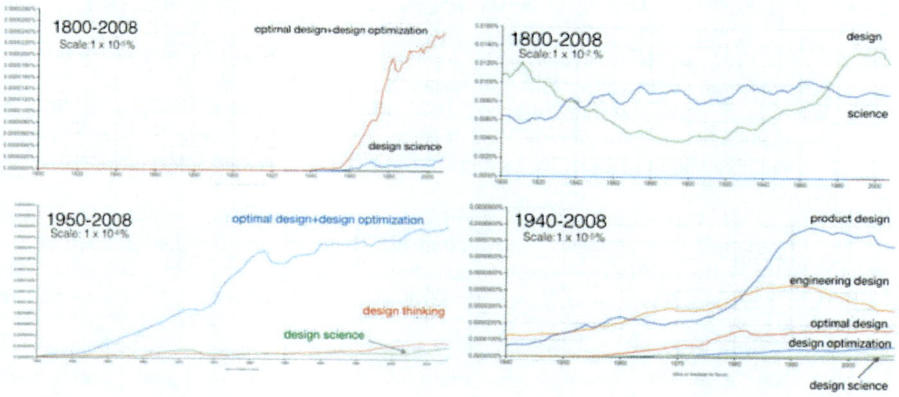

Fig. 2.9 Google Ngrams; frequency of terms occurring in books for the indicated time periods (accessed 25 September 2018). Note differences in scale

References

Alyaqout, S. (2006). *A Multi-System Optimization Approach To Coupling in Robust Design and Control*. PhD dissertation, Dept. of Mechanical Engineering, University of Michigan, Ann Arbor.

Bayrak, A. E., Kang, N., & Papalambros, P. Y. (2016). Decomposition based design optimization of hybrid electric powertrain architectures: Simultaneous configuration and sizing design. *Journal of Mechanical Design, 138*(7), 071405.

Bayrak, A. E., Collopy, A. X., Papalambros, P. Y., & Epureanu, B. I. (2018a). Multiobjective optimization of modular design concepts for a collection of interacting systems. *Structural and Multidisciplinary Optimization, 57*(1), 83–94. https://doi.org/10.1007/s00158-017-1872-4.

Bayrak, A. E., Egilmez, M. M., Kuang, H., Li, X., Park J. M., Umpfenbach, E., Anderson, E., Gorsich, D., Hu, S. J., Papalambros, P. Y., & Epureanu, B. I. (2018b). A system-of-systems approach to the strategic feasibility of modular vehicle fleets. *IEEE Transactions on Systems, Man, and Cybernetics*. https://doi.org/10.1109/TSMC.2018.2827387.

Bayrak, A. E. (2015). *Topology Considerations in Hybrid Electric Vehicle Powertrain Architecture Design*. PhD dissertation, Dept. of Mechanical Engineering, University of Michigan, Ann Arbor.

Bendsøe, M. P. (1995). *Optimization of Structural Topology, Shape, and Material*. Springer.

Bendsøe, M., & Kikuchi. (1988). Generating optimal topologies in structural design using a homogenization method. *Computer Methods in Applied Mechanics and Engineering, 71*, 197–224.

Brown, T. (2009). Design thinking. *Harvard Business Review, 86*, 84–92.

Burnap, A., Ren, Y., Gerth, R., Papzoglou, G., Gonzalez, R., & Papalambros, P. Y. (2015). When crowdsourcing fails: A study of expertise on crowdsourced design evaluation. *Journal of Mechanical Design, 137*(3), 031101. https://doi.org/10.1115/1.4029065

Burnap, A., Pan, Y., Liu, Y., Ren, Y., Lee, H., Gonzalez, R., & Papalambros, P. Y. (2016). Improving design preference prediction accuracy using feature learning. *Journal of Mechanical Design, 138*(7), 071404.

Chen, T., Mueller, J., & Shea, K. (2017). Integrated design and simulation of tunable, multi-state structures fabricated monolithically with multi-material 3D printing. *Scientific Reports, 7*, 45671. https://doi.org/10.1038/srep45671

Chen, W., Hoyle, C., & Wassenaar, H. J. (2012). *Decision-based Design: Integrating Consumer Preferences into Engineering Design*. Springer Science & Business Media.

Cross, N. (2011). *Design Thinking: Understanding How Designers Think and Work*. Berg Publishers Oxford.

Dede, E. M., Ll, J., & Nomura, T. (2014). *Multiphysics Simulation: Electromechanical System Applications and Optimization*. Springer-Verlag.

Ersal, T., Ahn, C., Peters, D. L., Whitefoot, J. W., Mechtenberg, A., Hiskens, I. A., Peng, H., Stefanopoulou, A. G., Papalambros, P. Y., & Stein, J. L. (2013). Coupling between component sizing and regulation capability in microgrids. *IEEE Transactions on Smart Grid, 4*(3), 1576–1585.

Fathy, H. K. (2003). *Combined Plant and Control Optimization: Theory, Strategies and Applications*. PhD dissertation, Dept. of Mechanical Engineering, University of Michigan, Ann Arbor.

Fleury, C. (1982). Reconciliation of mathematical programming and optimality criteria methods. In A. J. Morris (Ed.), *Foundations of Structural Optimization: A Unified Approach* (pp. 363–404). Wiley.

Gelles, D. (2018). *Tech Backlash Grows as Investors Press Apple to Act on Children's Use*. New York Times. https://www.nytimes.com/2018/01/08/technology/apple-tech-children-jana-calstrs.html. Accessed 25 Sept. 2018.

Georgiopoulos, P., Jonsson, M., & Papalambros, P. Y. (2005). Linking optimal design decisions to the theory of the firm: the case of resource allocation. *ASME Journal of Mechanical Design, 127*(2005), 358–366.

Hollister, S. J., Maddox, R. D., & Taboas, J. M. (2002). Optimal design and fabrication of scaffolds to mimic tissue properties and satisfy biological constraints. *Biomaterials, 23*(20), 4095–4103.

Kang, N., Ren, Y., Feinberg, F. M., & Papalambros, P. Y. (2016). Public investment and electric vehicle design: A model-based market analysis framework with application to a USA-china comparison study. *Design Science, 2*(e6), 1–42. https://doi.org/10.1017/dsj.2016.7

Kelly, J. C., Wakefield, G. H., & Papalambros, P. Y. (2010). Evidence for using interactive genetic algorithms in shape preference assessment. *International Journal of Product Development, 13*(2), 168–184.

Kelly, J. C., Maheut, P., Petiot, J. F., & Papalambros, P. Y. (2011). Incorporating user shape preference in engineering design optimization. *Journal of Engineering Design, 22*(9), 627–650.

Lewis, K., Chen, W., & Schmidt, L. (2006). (eds.) *Decision Making in Engineering Design*. ASME Press, New York.

Luk, J. M., Kim, H. C., De Kleine, R., Wallington, T.J., & MacLean, H. L. (2017). Review of the fuel saving, life cycle GHG emission, and ownership cost impacts of lightweighting vehicles with different powertrains. *Environmental Science & Technology, 51*(15), 8215—8228. https://doi.org/10.1021/acs.est.7b00909.

Michalek, J. J., Feinberg, F. M., & Papalambros, P. Y. (2005). Linking marketing and engineering product design decisions via analytical target cascading. *Journal of Product Innovation Management: Special Issue on Design and Marketing in New Product Development, 22*, 42–62.

Michalek, J. J., Ebbes, P., Adiguzel, F., Feinberg, F., & Papalambros, P. Y. (2010). Enhancing marketing with engineering: optimal product line design for heterogeneous markets. *International Journal of Research in Marketing, 28*(1), 1–12. https://doi.org/10.1016/j.ijresmar.2010.08.001

Michalek, J., Papalambros, P. Y.& Skerlos, S. (2004). A study of fuel efficiency and emission policy impact on optimal vehicle design decisions. *Journal of Mechanical Design, 126*(6), 1062–1070.

Murphy, S. V., & Atala, A. (2014). 3D Bioprinting of Tissues and Organs. *Nature Biotechnology, 32*, 773–785. https://doi.org/10.1038/nbt.2958

Papalambros, P. Y., & Georgiopoulos, P. (2006). In: Lewis, K., Schmidt, L., Chen, W., (eds.) *A Designer's View to Economics and Finance, in Decision Making in Engineering Design,* ASME Press, NY.

Papalambros, P. Y., & Wilde, D. J. (2017). *Principles of Optimal Design: Modeling and Computation* (3d ed.). Cambridge University Press.

Peters, D. L. (2010). *Coupling and Controllability in Optimal Design and Control.* PhD dissertation, Dept. of Mechanical Engineering, University of Michigan, Ann Arbor.

Reid, T., Gonzalez, R., & Papalambros, P. Y. (2010). Quantification of perceived environmental friendliness for vehicle silhouette design. *Journal of Mechanical Design, 132*(10), 101010.

Reid, T., Frischknecht, B., & Papalambros, P. Y. (2012). Perceptual attributes in product design: fuel economy and silhouette-based perceived environmental friendliness tradeoffs in automotive vehicle design. *Journal of Mechanical Design, 134*(4).

Ren, Y., Bayrak, A. E., & Papalambros, P. Y. (2016). EcoRacer: game-based optimal electric vehicle design and driver control using human players. *Journal of Mechanical Design, 138*(6), 061407.

Ren, Y., & Papalambros, P. Y. (2011). A design preference elicitation query as an optimization process, *Journal of Mechanical Design, 133*(11).

Ren, Y., & Papalambros P. Y. (2012). On design preference elicitation with crowd implicit feedback, In *Proceedings of the ASME 2012 International Design Engineering Technical Conferences*, Chicago, IL, DETC2012–70605.

Reyer, J. (2000). *Combined Embodiment Design and Control Optimization: Effects of Cross-Disciplinary Coupling.* PhD dissertation, Dept. of Mechanical Engineering, University of Michigan, Ann Arbor.

Siddall, J. N. (1972). *Analytical Decision-Making in Engineering Design.* Prentice Hall.

Simon, H.A. (1988). The Science of Design: Creating the Artificial. *Design Issues, 4*(1/2), *Special Issue on Designing the Immaterial Society*, pp. 67–82.

Wagner, T. C., & Papalambros, P. Y. (1999). Decomposition analysis and optimization of an automotive powertrain design model. *Engineering Optimization, 31*(3), 273–299.

Chapter 3
DESIGN, Building a Global, Interdisciplinary Community

Chris McMahon and Anja Maier

Abstract This paper reviews the development of the design research community over the thirty-seven years since the first scientific and professional meeting on "The Science of Design and Computer Aided Design" was held in Zagreb in 1981. It does this in the light of the Zeitgeist of the times, considering the developments broadly in four separate decades. For each decade, observations are made on the Zeitgeist, on the key defining events and characteristics of the era, and then the development at the time of the design research community is considered, exploring how it responded to the Zeitgeist, and picking out some key achievements. Key milestones in the development of that community will be examined, and reflections will include the impacts of different disciplines, theoretical perspectives and cultural influences that have arisen in the development of a worldwide community.

3.1 Introduction

When the first scientific and professional meeting on "The Science of Design and Computer Aided Design" was held in Zagreb in 1981 (Design, 1981), design practice, teaching and the then weak shoots of research were a very fragmented affair, with pockets of strong interest especially for example in Denmark, Germany and to some

C. McMahon
Department of Mechanical Engineering, Technical University of Denmark, Kongens Lyngby, Denmark

C. McMahon (✉)
Faculty of Engineering, University of Bristol, Bristol, United Kingdom
e-mail: chris.mcmahon@bris.ac.uk

A. Maier
Department of Design, Manufacture and Engineering Management, University of Strathclyde, Glasgow, United Kingdom

Department of Technology, Management and Economics, Technical University of Denmark, Kongens Lyngby, Denmark

D. Marjanović et al. (eds.), *Design Research: The Sociotechnical Aspects of Quality, Creativity, and Innovation*, https://doi.org/10.1007/978-3-031-50488-4_3

extent the UK, yet no general recognition of the subject as of particular interest for academic enhancement through research.

The scene was, however, set for the great developments we have seen in the past 37 years. Many countries had been scarred by fuel crises and stagflation in the challenging years of the 1970s. Japan's economic miracle had stunned the world, with its emphasis on quality and efficiency in production and on the learning organisation (Smith, 1995). Deng Xiaoping had taken power in China in 1978, and was instituting significant economic reforms (Vogel, 2013). The elections of Reagan in the USA, Thatcher in the UK and Kohl in Germany reflected what became the strong neo-liberal and globalising agenda of the end of the twentieth century, soon to be strengthened by the collapse of the Soviet Union and the opening up of Eastern Europe and the People's Republic of China. The world was ready for a discipline whose focus was fundamentally on the creation and improvement of quality products, on creativity and innovation, on invention and productivity.

The design research community was ready to build on the developments that have been made by pioneers in systematic design and in design theory and methodology, on strong practices and experience in industry and also on the emerging interest in companies and the academic world in computer-aided design, artificial intelligence and other applications of information technologies in engineering. Of increasing interest also were the socio-technical aspects. Without design engineering, science would stay in the lab. Design research is intrinsically linked with design practice and design methodology is as much about technical design rules as about their socio-technical implementation. The developing social sciences were also to prove a strong driver (Cross, 2001).

This paper will present some perspectives on the way design research has developed in the intervening years since the very Zagreb meeting, and especially the way the European interest reflected in the DESIGN series of conferences has formed a core on which a community has been built. It is nearly four decades—a working lifetime—and as we considered the events of those years, those decades are used to organise our overall approach to the presentation as follows. For each decade, observations will be made on the *Zeitgeist*, on the key defining events and characteristics of the era. Remarks will then be made about the development of the design research community, and how (if at all) we feel it responded to the *Zeitgeist*, again picking out some key achievements of the time (of course we can only give examples, and we miss many). We reflect on the influence of different disciplines, on the theoretical perspectives and cultural influences that have come from a worldwide community that has developed—thanks substantially to the work done by those who have organised the professional communities and the DESIGN and other conferences—from a fragmented to a global community. We will conclude with the observation that the DESIGN conference series has proved a perfect forum for the community that has emerged, presenting as it does a balance between industrial significance and scholarly rigour, a welcome for those with an interest in design from many disciplines and a world heritage site at which to meet.

3.2 The 1980s—Foundation

We begin by casting our minds back to the 1980s. As noted, we feel that we cannot consider the 1980s without considering what they followed—what some people have called the 'traumatic 1970s' (Keys, Davies & Bannan, 2014). It was of course a time of great scientific and cultural development, with strong achievements in the performing arts, but in political terms it was certainly traumatic, with the end of what the French call the *trente glourieuses* (Fourastié, 1979) —the thirty glorious years of prosperity and growth—stagflation (the combination of economic stagnation and inflation), and the fall of the Bretton Woods system of international monetary management, all associated with the oil crisis of the time (Hammes & Wills, 2005). A reaction to those events was the election of Ronald Reagan in the USA and Margaret Thatcher in the UK—and together with the election of Helmut Kohl in West Germany and the elevation of Deng Xiaoping as 'paramount leader' of the People's Republic of China (and others) this helped to instigate what became the strong neo-liberal and globalising agenda of the end of the twentieth century.

In addition to the geopolitical events of the 1960s and 1970s, the period had also seen the dramatic rise of Japanese manufacturing, with the country's manufacturers developing leading positions especially in automobile manufacture, consumer electronics and ship building. In part, this was aided by the oil crises leading to increased demand for small automobiles, but foremost were the advances on productivity, quality and process efficiency made by Japanese manufacturers (Cusumano, 1988). As Womack and colleagues said the Japanese 'machine' truly changed the world in the 1970s and 80 s, with just-in-time, and lean, learning organisations (Womack et al., 1990).

And of course, the beginning of the 1980s were the great turning point for computing—the VAX 11/780 and Data General MV8000 minicomputers were making significant computing power available for the first time to many companies (Kidder, 1981). The Motorola 68,000 from 1979 (which was the heart of many early engineering workstations) and then the IBM PC of 1981 (Malone, 2014) provided the real basis for ubiquitous computing and the great development of many informatics applications (McMahon, Liu & McAdams, 2016). At the same time consumer electronics was transforming products, with the release of the Sony Walkman in 1979 allowing people to listen to music on the move (Du Gay et al. 1997), and the introduction of a number of video games at about the same time (Kent, 2001). These developments would have benefits for CAD, owing to the introduction of low-cost graphics capabilities for gaming.

3.2.1 Design Research Emerging

So, in the 1980s, the world was truly ready for a discipline whose focus was on the creation and improvement of quality products, on creativity and innovation, on invention and productivity and on how the engineering world could use the opportunities that computing offered. Engineering design research could build on a whole range of existing contributions: systematic design research in Germany, Czechoslovakia and Denmark (Bayazit, 2004), a long tradition of work in machine elements (Pahl et al., 1999), the design methods movement that had been especially strong in the UK in the 1960s and 70s (Bayazit, 2004), design automation and optimisation especially in the USA (Ragsdell et al., 2015), widespread work in applied art and design. And of course, a great deal of celebrated work had already been published – by numerous researchers in the German school and by the likes of Vladimir Hubka, Mogens Andreasen, Christopher Jones, Christopher Alexander, Morris Asimow, Bruce Archer and Herbert Simon. Many techniques of the Toyota Production System were in place, including practices like single minute exchange of dies, QFD (originally introduced by Mitsubishi) and so on (Chiarini, 2013). The main computational tools of design—geometric modelling, finite element analysis, computational fluid dynamics, multibody dynamic analysis were all in place (McMahon et al., 2016). And researchers had venues for discussion and dissemination of their work in the Design Research Society (founded in 1966), the American Society of Mechanical Engineers Journal of Mechanical Design (ASME JMD, 1978) and Design Automation Conference (DAC, 1976), and the Design Studies (1979) and Computer-aided Design journals (1968), among others—but these were scattered shoots, each covering only part of the community's work. A coherent forum was needed.

At the time of the Zagreb conference the other major development in Europe was the formation of Workshop Design Konstruktion (WDK) and the holding of the first International Conference on Engineering Design (ICED) in Rome (Hubka, 1981), the first of 13 under the auspices of WDK, ending in Glasgow in 2001. Research funders were waking up to the possibilities of research in engineering design – in the UK, the Science and Engineering Research Council (SERC) had an 'Application of Computers to Manufacturing Engineering' (ACME) directorate with a theme on CAD of the product and its means of production (Smith, 1986), and the US National Science Foundation had a programme of research into design methodology. The English translation by Ken Wallace of Pahl and Beitz's Konstruktionslehre (Pahl & Beitz, 1984) was a landmark of the 1980s (as was Crispin Hales PhD in design, the first at Cambridge (Hales, 1987)), and other important books at that time included Hubka and Eder's Principles of Engineering Design (Hubka & Eder, 1982), French's Conceptual Design (French, 1985), Cross's Engineering Design Methods (Cross, 1989) and Schön's Reflective Practitioner (Schön, 1983) among others.

What was the focus of the community at that time? At the beginning of the twenty-first century Mogens Andreasen, reflecting on the WDK ICED conferences, identified four key themes (Andreasen, 2001). The first was a large group of papers seeking to articulate what constituted design science: what was the scientific basis for the subject and what were the appropriate research methodologies to be used in its study? The second reflected a development from a concentration on mechanical design, especially machine design, to a wider emphasis on product development. There was in particular a developing interest in 'design for X' (DfX), where 'X' described life cycle properties of the designed artefact that included especially manufacturability and assemblability but also issues related to environmental performance, and to design for the whole life cycle. This group also contained papers on team work, on the human aspects of design—including collaboration and creativity. The third large group of papers reflected the strong interest at the time in CAD, but also showed developing emphasis on wider application of information technologies in many aspects of design from synthesis to information and knowledge management and many aspects of modelling. A fourth and final group of papers Andreasen entitled "delimitations of ICED", describing the papers in this group as broadening out from the engineering focus to a wider interest in innovation more generally The emphasis on product development reflected very strongly the competitive industrial culture of the time, and design for X—design for manufacture in particular—was perhaps a response to the rapidly escalating labour costs of the 1970s and the prowess that the Japanese had shown in manufacturing productivity, but also to the increasing awareness of environmental issues. The emphasis on information technology was a natural product of the rapidly developing computing technologies, and in particular the transition to CAD in industry which was taking place very strongly at the time.

3.3 1990s—Growth

Writing from a British and German perspective, two of the iconic images of the end of the 1980s were British comedian Harry Enfield's 'Loadsamoney' character—who boasted about his wealth by waving a wad of cash, and who was seen by many as emblematic of the brash materialism of Margaret Thatcher's Britain of the time—and the fall of the Berlin Wall, symbolising as it did the end of the post-1945 world. It has been suggested that the two are related. The North Sea and Alaska North Slope had for a time assuaged concerns about oil supplies, and led to low oil prices, but these damaged the oil-export dependent Soviet economy (Thompson, 2017). These issues reflected the political *Zeitgeist* of the time, which was one of strong growth, economic liberalism and geopolitical change.

The events at the turn of the decade were sufficiently dramatic that Francis Fukayama had the temerity to write in 1989 of the 'end of history' (Fukuyama, 1989) and certainly the fall of the Soviet Union, German reunification, the creation of the European Union with the Maastricht Treaty, its subsequent eastward expansion but also the return of Hong Kong to the People's Republic of China and the continued opening of China were the making of our modern world order. Though, as we know, these were also troubled and painful times of transition. The geopolitical changes were accompanied by continuing development of the neoliberal agenda—financial deregulation and, for many, economic prosperity, fueled in part by the transformative technologies of the Internet and by then the World Wide Web, by mobile telephony becoming a mass consumption product and by microprocessors with everything.

3.3.1 A Developing Design Community

In design research we saw a continuation of the trends established in the 1980s, with strong growth driven by similar factors to those that prevailed in the 1980s. Attendance at ICED conferences grew to more than 600 at Munich in 1999, the Design Theory and Methodology (DTM) conference had been started by ASME in 1989, in Croatia, the Zagreb conference morphed into DESIGN, and in 1991 the International Academy for Production Engineering (CIRP) introduced a design conference. It was an especially fruitful time for design journals, with the creation of Research in Engineering Design (RED), the Journal of Engineering Design (JED) and also AI in Engineering Design and Manufacture (AIEDAM), and Advanced Engineering Informatics (at that time 'AI in Engineering'), the latter two reflecting the importance of informatics in the discipline. There was a real feeling of design research becoming mainstream—it could be found, for example, at many of the leading technical universities of the world, and there were a number of engineering design research centres established, in particular at Carnegie Mellon University in the USA (Talukdar, 2001) and the University of Cambridge in the UK.

It was also a time when there was a great sharing of ideas and a mixing of perspectives. Theoretical perspectives on design were expanded, with significant contributions from Australia, Japan and the USA extending the rich set of outlooks from Europe (Chakrabarti and Blessing, 2015). The industrial, applications-based focus of some parts of Europe met the more laboratory-based and experimentally-focused cultures of other communities, and there was particular interest in research methodology, with learning from other scientific communities including psychology, management studies and computing. There was a strong melding of the pragmatic with the theoretical and the systematic. As always in design research, industrial engagement and collaboration was important, but this was also a time of growing emphasis on journal publication, and on establishing the academic credentials of the community's work. In the research itself, there was particular emphasis on product improvement, on customer focus, on tools and methods (esp. computational) for industry, with increasing diversity—for example with growing interest in creativity, in human behaviour, in collaboration, in risk and uncertainty and in systems approaches (The Design Society, 2018).

3.4 2000s—Consolidation

We have used the term 'consolidation' for the 2000s, and that is true for the design research community. The *Zeitgeist* has been somewhat more mixed, with a turbulent decade whose effects are still being felt very strongly.

The early years of the 2000s were quite remarkable, with continued growth in world trade, the growing importance of Brazil, Russia, India, China and South Africa (the BRICS), continued expansion of the European Union and the introduction of the Euro currency. Global brands were very important, with names like Coca-Cola, Microsoft, IBM, GE, Toyota and Intel dominant (Interbrand, 2007). Nokia's position at number 5 reflected the importance of mobile communication, but its fall, and the growth of the new 'technology' brands (in 2017 five of the top eight positions would be held by Apple, Google, Samsung, Amazon and Facebook (Interbrand, 2017)) reflected how dynamic the technology market was at that time. The early years of the decade were particularly turbulent, with the bursting of the dot-com bubble in 2001, although the bubble had at least led to the building of the infrastructure of the world wide web (Janeway, 2012), had allowed all the foundational technologies to be developed (very often with public funding) (Mazzucato, 2011) and had nurtured some companies that knew how to make money from it.

However, a growing world also meant growing impacts. The shocks of the Twin Towers attacks on September 11, 2001 and the global financial crisis of 2007–8 have strongly coloured the years that followed. These years have also been characterised by growing awareness of the impact of our voracious appetite for materials and energy and the environmental damage that consumption causes, with the price of oil reaching a peak of over $140 per barrel in 2008 as (many argued) conventional oil production reached its highest ever levels (Kerr, 2011), and the Kyoto Protocol for the control of greenhouse gas emissions coming into force in 2005 (Grubb et al., 1999).

3.4.1 The Design Society and Cavtat

At the beginning of the 2000s the DESIGN Conference was held in its home for the next 16 years, the Hotel Croatia in Cavtat; at about the same time as the founding of the Design Society in 2001, taking over the duties of WDK and organising the ICED conference series from ICED 2003 when it was held in Stockholm onwards (Folkeson et al. 2003). The Design Society continued the work of WDK and expanded it through a programme of special interest groups (SIGs) and chapters, these themselves initiating new conferences including the International Conference on Design Creativity and the Engineering and Product Design Education conference, also in due course linking through endorsement with established events like DESIGN and also the International Conference on Research in Design (ICoRD), held in India, the Dependency and Structure Modelling (DSM) conference, and so on.

WDK had held an ICED conference in Boston in 1987. This had been the only one outside of Europe under WDK direction. In 2005, the DS began the practice of holding every second conference out of Europe, with events in Australia in 2005 and California in 2009. As well as the more international outlook, the Society introduced expanded papers—extending the number of pages from 4 or 6 in WDK days to 10—and unified conference themes and organisation. From 2009 the proceedings were published electronically and on paper, with one volume corresponding to each theme. The balance of the papers in each theme is a good indication of the interests of the research community in this decade: Table 3.1 shows the themes and the proportion of papers from the 2009 conference in each theme. The continuing strength of interest in design methods and tools is clear, as is the emphasis on 'design for X' and on 'design theory and research methodology' (which developed particularly strongly in that decade, with the establishment of a Design Theory SIG in 2008, led by Mines Paristech in Paris, and the publication of Blessing and Chakrabarti's influential DRM book in 2009 (Blessing and Chakrabarti, 2009)). Topics that had developed strongly in the 2000s were 'design processes' and 'design organisation and management', and especially 'human behavior in design', which by 2013 would encompass about 17% of the published papers and which had a strong emphasis on topics such as creativity, collaboration, risk perception and the emerging topic of design for emotion. Perhaps surprisingly, CAD was not a separate topic, although 'design information and knowledge' had a strong emphasis on the use of informatics (Table 3.1)

By 2009, the ICED conferences were indexed in the Web of Science and Scopus. In even numbered years, DESIGN, also by now indexed, became a central feature of the design research calendar. The interest in design creativity led to the establishment of an International Conference in Design Creativity, held for the first time in Kobe, Japan in 2010. The Engineering and Product Design Education (EPDE) conference, which had been held in the UK since 1999, broadened to the continent of Europe with very successful conferences being held in Delft in 2004, Salzburg in 2006 and Barcelona in 2008. More broadly than the DS, and perhaps a mixed blessing, as it

Table 3.1 Proportion of Papers in ICED2009 by topic

Theme	Proportion of papers %
Design processes	14
Design theory/research methodology	11
Design organisation and management	9
Product, service and system design	8
Design methods and tools	23
Design for X, design to X	8
Design information and knowledge	9
Human behaviour in design	9
Design education and lifelong learning	8

spread a particular and rather narrow view of design, the idea of 'design thinking' began to be widely popularised in business processes (Brown, 2009).

3.5 2010s—Reflection and Interdisciplinarity

And so, we come to the present decade, and again the political and social *Zeitgeist* and that of the design research community appear to have quite different emphases.

Globally, it has again been a time of uncertainty, with the continuing effects of the global financial crisis and a time of change in global and local politics: Brexit, Trump, the Eurozone crisis, migration and immigration, civil war in Syria. Our environmental concerns continue, with the International Panel on Climate Change advising us of the changes in our fossil fuel consumption habits that we need if we are to avoid catastrophic climate change (Drouet et al., 2015). Even some of the gloss has come off our view of Big Tech, with concerns about 'fake news', cyber-attacks, the polarising effect of social media and the addictive effects of mobile phones (Dredge, 2018; Prime Minister's Office and Finland, 2018).

High-tech has also contributed to a culture everywhere of ratings and ranking, with the consequent constant pressure on individuals and organisations to improve their rankings—the 'tyranny' of the h-index and the impact factor (Colquhoun, 2003). For both individuals and communities this can lead to pressure to 'improve' their ratings and scientific legitimacy, which can mean pressure to conform to particular views on what is correct or appropriate in scientific terms, and reduction in diversity and adventure in research.

3.5.1 The Response of the Design Community

The design research community has responded strongly to the ratings and ranking pressures, building a more professional community, in particular under the leadership of the Design Society, with the new Design Science Journal (Papalambros, 2015), open access for all the community's papers, continued emphasis on research quality and a well re-engineered web site. In terms of scientific outputs there has been strong recent production in books (Chen et al., 2013; Myrup et al., 2015; Chakrabarti & Lindemann, 2016; Le et al., 2017) and the DS website now has many thousands of papers on-line and in open access (The Design Society, 2018). Concerns about the strength of the community's research remain, with active articulation and discussion of approaches to improve its robustness and impact (Cash, 2018).

There has also been a continual growing of the community's interests beyond the original emphasis on machines and products—with emphasis on engineered systems, a greater interest in industrial design, a growth of work on servitisation and on the combination of products and services in product-service systems. The continuing importance of information technologies is shown less in their utilisation in design tools and methods (although these are still active topics in the community) than in their exploitation in the designed artefact—in human–computer-interaction, in design for an ubiquitously interconnected world. And there is continuing interest in how we achieve novelty, in the management of innovation and entrepreneurship, in part again because of the need to reinvent for sustainability, but also reflecting the search for new products and systems in the light of stagnating economies and new approaches to meet global development goals (United Nations, no date; Szirmai et al., 2011).

All of this reflects the state of society—its growing interconnectedness, the growing importance of services in the advanced economies and in reducing environmental impact. It also means that the design research community is becoming less homogeneous, and perhaps less easy for people to find their place. We need to reflect what our core interests are, and on where we are placed with respect to other research communities.

3.6 Now and Going Forward

In the past four decades the design research has built a robust body of knowledge and a thriving international community at which the DESIGN Conference and WDK then the Design Society have been at the heart. This community has been built on strong links to industry and has grown out of foundations in mechanical engineering and product design. It has in its development actively responded to the needs of industry and the economy, but arguably it has also been rather reactive, responding rather than leading. At a time of uncertainty in multiple dimensions—in the economy, in national and international politics, in humanity's impact on and relationship with the Earth—there is a need for research that will help us change those aspects of the designed, artificial world that are intimately connected with these issues. And there is also a need for research that will help us to understand and adapt the complex, 'engineered systems' that make up much of that artificial world (deWeck et al., 2011). In these regards, what can the design research community offer?

3.6.1 Research Methods for the Future

In the past forty years we have concentrated, for example on the design of products, on the development of tools to assist the designer, on methods to improve the performance of the designed artefact, including computational approaches, and on understanding the human behaviour of the designer, of design teams and customers. In the newly challenging world, we need research methods that allow us to understand the complex nature of our socio-technical systems, but also to be able to propose and develop radical solutions to the challenges that are faced. Some thoughts on the design research approaches that might assist this include:

- Approaches that allow us to study 'design as a complex fabric of interwoven processes' (Cash et al., 2015) at multiple scales, especially those for system modelling and simulation, but also for system observation and design of experiments at a system level.
- Techniques that allow us to take advantage of the vast quantities of data that can be gathered in all aspects of the creation and operation of our engineered systems: data science, 'Big Data', 'Thick Data' (Wang, 2013), but also live data, sensor data for condition monitoring, observations and predictions, and monitoring of technical and human behavioural patterns (Thorpe et al., 2017).
- Methods and approaches that have a stronger engagement with the wider community: living labs, citizen science and crowd research (Kareborn and Stahlbrost, 2009; Salganik, 2018).

The key in this will be not simply to conceive and then improve new products, but rather to reconceive the system basis for the engineered world, especially one that is better aligned to the true carrying capacity of the planet. The approaches that we use may in addition be used in the redesign of systems in which engineering design has traditionally had a minor role—including, e.g. healthcare, legal systems, and policy.

3.6.2 New Theoretical Foundations

Our overall aim must be to identify and then actively pursue an active role for the future of humanity. Design research should articulate more strongly what it can offer and do well, especially in tackling the societal challenges that we have noted. In addition to needing new methodological approaches, such research will also need to build on and expand that theoretical foundations that have been established over the past decades. Again, the key here may well be to enhance the system perspective, in health, in energy, in transportation, in agriculture, in manufacturing, in policy, and this is likely to mean a melding of theoretical perspectives from design and systems (Maier et al., 2022; McMahon & Krumdieck, 2022). This will show design as a connector between fields and will also provide a more robust basis for linking of process to performance outcomes ('success science').

The theoretical foundations may also be extended by building on design research in decision-making and again extending it to a system framework. Tools for modelling and visualising of systems may be enhanced through reasoning and decision-support to engage in the objectives of reflective, question-driven and data-supported human intelligence.

3.7 Conclusions

So how might we summarise? We started forty years ago from geographically fragmented beginnings, but substantially in the design of ‘mechanical’ products - in machine design, in design methods, in design automation, in CAD. Industry had majored in ‘lean’ and quality management. The focus of research was still mainly through local communities, with relatively little industrial and funding body support, and no real tradition of academic research. There were few if any PhDs in design, and few academic posts with ‘design’ in their titles. But there were fresh shoots and some real enthusiasm in some pioneer institutions and from industry. Research centred on understanding machines, and the methods and tools that could be used in their design, especially using the rapidly developing computing technologies.

Developments in Zagreb in 1981 were among the fresh shoots which led to a pattern of strong growth in engineering design research at the end of the twentieth century, establishing a vibrant community, with a sound scientific foundation, especially through the work of WDK, the ASME and the Design Research Society. That growth coincided with an industrial transformation—enormous development through digitalisation and globalisation, with a strong emphasis on quality and the customer that have led to the formidable products and systems of the industrial world. Design research has contributed significantly to the excellence of these products and systems.

In the second decade of the twentieth century we are in a time of uncertainty and doubt. The global financial crisis has led to economic stagnation and challenges to globalisation. We are more aware than ever of the externalities of the industrial world—of environmental damage and on the social impacts of technical change. We are questioning of the direction of technology and society and that questioning should be reflected in the discussions on the future of the design research community. The research has always reflected the *Zeitgeist* and in the twentieth century was very much driven by national and industrial needs. In an uncertain world, perhaps it is time for design research to actively articulate more strongly what it can offer and do well, especially in tackling global and societal challenges. Perhaps, also, it is time to give the choice and crafting of research questions particular focus? In those regards, we hope that the DESIGN conference will continue to be the perfect forum for the community, continuing its balance between industrial significance and scholarly rigour, while providing an environment for the community to develop and nurture its future direction.

References

Andreasen, M. M. (2001) ‘The contribution of design research to industry - reflections on 20 years of ICED conferences’. In *Proceedings of the International Conference on Engineering Design, ICED01*. London: Professional Engineering Publishing.

Bayazit, N. (2004). 'Investigating design: A review of forty years of design research'. *Design Issues, 20*(1), 16–29. MIT Press 238 Main St., Suite 500, Cambridge, MA 02142–1046 USA journals-info@mit.edu. doi: https://doi.org/10.1162/074793604772933739.

Blessing, L., & Chakrabarti, A. (2009). *DRM: A Design Reseach Methodology*. Blessing, L., & Chakrabarti, A (eds.). London: Springer London. https://doi.org/10.1007/978-1-84882-587-1_2.

Brown, T. (2009). Change by design: how design thinking transforms organizations and inspires innovation.

Cash, P., Hicks, B., & Culley, S. (2015). Activity Theory as a means for multi-scale analysis of the engineering design process: A protocol study of design in practice. *Design Studies, 38*, 1–32. https://doi.org/10.1016/j.destud.2015.02.001

Cash, P. J. (2018). Developing theory-driven design research. *Design Studies, 56*, 84–119. https://doi.org/10.1016/j.destud.2018.03.002

Chakrabarti, A., & Blessing, L. (2015). 'A Review of Theories and Models of Design', *Journal of the Indian Institute of Science, 95*(4), 325–340.

Chakrabarti, A. & Lindemann, U. (2016). *Impact of Design Research on Industrial Practice: Tools, Technology, and Training*. In: Chakrabarti, A., & Lindemann, U. (eds.) Cham: Springer International Publishing. doi: https://doi.org/10.1007/978-3-319-19449-3.

Chen, W., Hoyle, C., & Wassenaar, H. J. (2013). *Decision-Based Design: Integrating Consumer Preferences into Engineering Design*. London: Springer, London. https://doi.org/10.1007/978-1-4471-4036-8_10

Chiarini, A. (2013). *Lean organization: From the tools of the toyota production system to lean office*. Springer.

Colquhoun, D. (2003). Challenging the tyranny of impact factors. *Nature, 423*(6939), 479–479. https://doi.org/10.1038/423479a

Cross, N. (1989). *Engineering Design Methods*. Chichester: J. Wiley. Available at: https://www.wiley.com/en-us/Engineering+Design+Methods%3A+Strategies+for+Product+Design%2C+4th+Edition-p-9780470519264 (Accessed: 14 November 2018).

Cross, N. (2001). Designerly ways of knowing: Design discipline versus design science. *Design Issues. The MIT Press, 17*(3), 49–55. https://doi.org/10.2307/1511801

Cusumano, M. A. (1988). 'Manufacturing innovation: lessons from the japanese auto industry'. *MIT Sloan Management Review, 30*(1). Available at: https://sloanreview.mit.edu/article/manufacturing-innovation-lessons-from-the-japanese-auto-industry/ (Accessed: 13 November 2018).

Design 1981. (1981). DESIGN 1981—The science of design and computer aided design. In *DESIGN 2018 / DESIGN Conference*. Available at: http://www.designconference.org/past-events (Accessed: 13 November 2018).

deWeck, O. L., Roos, D., & Magee, C. L. (2011). *Engineering Systems: Meeting Human Needs in a Complex Technological World*. Boston, MA: MIT Press.

Dredge, S. (2018). Mobile phone addiction? *It's time to take back control | Technology | The Guardian*. The Guardian Newspaper. Available at: https://www.theguardian.com/technology/2018/jan/27/mobile-phone-addiction-apps-break-the-habit-take-back-control (Accessed: 13 November 2018).

Drouet, L., Bosetti, V., & Tavoni, M. (2015). Selection of climate policies under the uncertainties in the fifth assessment report of the IPCC. *Nature Climate Change*. https://doi.org/10.1038/nclimate2721

Du Gay, P. et al. (1997). *Doing cultural studies: The story of the Sony Walkman*. London: SAGE Publications in assiciation with The Open University. Available at: https://s3.amazonaws.com/academia.edu.documents/34594265/The_Story_of_the_Sony_Walkman.pdf?AWSAccessKeyId=AKIAIWOWYYGZ2Y53UL3A&Expires=1542176971&Signature=Lfc1YX6JR5zzs9%2BWx0mpLHh%2BdpY%3D&response-content-disposition=inline%3B filename%3DThe_Story_of (Accessed: 14 November 2018).

Folkeson, A. et al. (2003). '14th International Conference on Engineering Design (ICED03)', In *DS 31: Proceedings of ICED 03, the 14th International Conference on Engineering Design*.

Stockholm. Design Society, p. 701. Available at: https://www.designsociety.org/publication/38/DS+31%3A+Proceedings+of+ICED+03%2C+the+14th+International+Conference+on+Engineering+Design%2C+Stockholm (Accessed: 14 November 2018).

Fourastié, J. (1979). *Les Trente Glorieuses: Ou La Révolution Invisible de 1946 à 1975*. Paris: Fayard.

French, M. J. (1985). *Conceptual Design for Engineers*. Berlin, Heidelberg: Springer Berlin Heidelberg. doi: https://doi.org/10.1007/978-3-662-11364-6_1.

Fukuyama, F. (1989). 'The End of History?', *The National Interest, 16*, 3–18. Available at: https://www.embl.de/aboutus/science_society/discussion/discussion_2006/ref1-22june06.pdf (Accessed: 13 November 2018).

Grubb, M. et al. (1999). The Kyoto protocol: a guide and assessment. *Royal Institute of International Affairs, Energy and Environmental Programme*.

Hales, C. (1987). *Analysis of the Engineering Design Process in an Industrial Context*. University of Cambridge.

Hammes, D., & Wills, D. (2005). 'Black gold: The end of Bretton woods and the oil-price shocks of the 1970s: The independent review: the independent institute'. *The Independent Review: A Journal of Political Economy, 9*(4), 501–511. Available at: http://www.independent.org/publications/tir/article.asp?id=518 (Accessed: 13 November 2018).

Hubka, V. (ed.). (1981). 'WDK5 Design Methodology and WDK6 Teaching Engineering', In *Proceedings of ICED81. Rome: Workshop Design Konstruktion (WDK)*. Available at: https://www.abebooks.com/book-search/author/hubka-v/first-edition/ (Accessed: 14 November 2018).

Hubka, V. (1982). Principles of Engineering Design (Translated and edited by W E Eder). London: Butterworth Scientific.

Interbrand (2007) Rankings—2007—Best global brands—Best brands—Interbrand, Interbrand. Available at: https://www.interbrand.com/best-brands/best-global-brands/2007/ranking/ (Accessed: 13 November 2018).

Interbrand. (2017). Rankings—2017—best global brands—best brands—Interbrand, Interbrand. Available at: https://www.interbrand.com/best-brands/best-global-brands/2017/ranking/ (Accessed: 13 November 2018).

Janeway, W. H. (2012). *Doing Capitalism in the Innovation Economy: Markets*. Speculation and the State: Cambridge University Press.

Kareborn, B. B., & Stahlbrost, A. (2009). Living Lab: an open and citizen-centric approach for innovation. *International Journal of Innovation and Regional Development, 1*(4), 356. https://doi.org/10.1504/IJIRD.2009.022727

Kent, S. L. (2001). The ultimate history of video games: From Pong to Pokemon. *The Story Behind the Craze That Touched Our Lives and Changed the World*. Prima Pub.

Kerr, R. A. (2011). Peak Oil Production May Already Be Here. *Science, 331*(6024), 1510–1511. https://doi.org/10.1126/science.331.6024.1510

Keys, B., Davies, J., & Bannan, E. (2014). 'the post-traumatic decade: New histories of the 1970s', Australasian journal of American studies. *Australia New Zealand American Studies Association, 33*(1), 1–17. https://doi.org/10.2307/44706134

Kidder, T. (1981). *The Soul of a New Machine*. Columbus, GA: Little, Brown and Company.

Maier, A., Oehmen, J., & Vermaas, P. E. (2022). Introducing engineering systems design: A new engineering perspective on the challenges of our times. In Maier, A., Oehmen, J., & Vermaas, P. E. (Eds.), *Handbook of Engineering Systems Design* (pp. 3–32). Springer. https://doi.org/10.1007/978-3-030-46054-9_1-2

Le Masson, P., Weil, B., & Hatchuel, A. (2017). *Design Theory: Methods and Organization for Innovation*. Cham: Springer International Publishing. https://doi.org/10.1007/978-3-319-50277-9

Malone, M. S. (2014). *The Intel Trinity: How Robert Noyce, Gordon Moore, and Andy Grove Built the World's Most Important Company*. New York: Harper Collins.

Mazzucato, M. (2011). *The Entrepreneurial State: Debunking Public versus Private Sector Myths*. London: Anthem Press.

McMahon, C., Krumdieck, S. (2022). Transitioning to sustainable engineering systems. In: Maier, A., Oehmen, J., & Vermaas, P. E. (Eds.), *Handbook of Engineering Systems Design*. Cham: Springer. https://doi.org/10.1007/978-3-030-81159-4_37

McMahon, C., Liu, Y., & McAdams, D. (2016). Engineering design informatics. *Artificial Intelligence for Engineering Design, Analysis and Manufacturing, 30*(04), 335–336. https://doi.org/10.1017/S0890060416000329

Myrup Andreasen, M., Hansen, C. T., & Cash, P. (2015). *Conceptual Design: Interpretation, Mindset and Models*. Cham: Springer International Publishing. https://doi.org/10.1007/978-3-319-19839-2_1

Pahl, G., Badke-Schaub, P., & Frankenberger, E. (1999). 'Resume of 12 years interdisciplinary empirical studies of engineering design in Germany'. *Design Studies, 20*(5), 481–494. Available at: http://www.sciencedirect.com/science/article/B6V2K-3X28Y3X-9/2/9a8095cd44638eef3fde17a1d7c15ea0.

Pahl, G., & Beitz, W. (1984). *Engineering Design: A Systematic Approach*. In: Wallace, K. (ed.), London: Springer London. https://doi.org/10.1007/978-1-4471-3581-4_1.

Papalambros, P. Y. (2015). Design Science: Why, What and How. *Design Science. Cambridge University Press, 1*, e1. https://doi.org/10.1017/dsj.2015.1

Prime Minister's Office and Finland. (2018). Countering hybrid threats: When fake news and cyber-attacks exacerbate crises. *7th Workshop on Strategic Crisis Management—OECD*. Available at: http://www.oecd.org/gov/risk/7th-workshop-strategic-crisis-management.htm (Accessed: 14 November 2018).

Ragsdell, K. et al. (2015). 'New Perspectives on Design Automation: Celebrating the 40th Anniversary of the ASME Design Automation Conference'. *Journal of Mechanical Design. American Society of Mechanical Engineers(ASME), 137*(5), 050301. https://doi.org/10.1115/1.4030256.

Salganik, M. J. (2018). *Bit by Bit: Social Research in the Digital Age*. Princeton: Princeton University Press.

Schön, D. A. (1983). *The reflective practitioner : How Professionals Think in action*. Basic Books.

Smith, D. B. (1995). *Japan Since 1945: The Rise of an Economic Superpower* (5th ed.). New York: Macmillan.

Smith, P. C. L. (1986). Application of computers to manufacturing engineering—a new SERC initiative. *Education and Training in Robotics* (pp. 209–215). New York: Springer.

Szirmai, A., Naudé, W., & Goedhuys, M. (2011). *Entrepreneurship, Innovation, and Economic Development*. Oxford University Press.

Talukdar, S. (2001). The engineering design research center: some lessons from its history. Vancouver: I E E E (*IEEE Power Engineering Society Summer Meeting*). Available at: https://www.tib.eu/en/search/id/BLCP%3ACN042480617/The-Engineering-Design-Research-Center-Some-Lessons/ (Accessed: 14 November 2018).

The Design Society. (2018). *Search DS publications/The Design Society*. Available at: https://www.designsociety.org/group/18/Search+DS+publications (Accessed: 13 November 2018).

Thompson, H. (2017). *Oil and the Western Economic Crisis*. Cham: Springer International Publishing. https://doi.org/10.1007/978-3-319-52509-9_1

Thorpe, J. R., Forchhammer, B. H.& Maier, A. (2017). 'Sensing behaviour in healthcare design', In *Proceedings of the International Conference on Engineering Design, ICED*.

United Nations (no date) *United Nations Sustainable Development Goals*. Available at: https://www.un.org/sustainabledevelopment/sustainable-development-goals/ (Accessed: 13 November 2018).

Vogel, E. F. (2013). *Deng Xiaoping and the Transformation of China*. Cambridge, Massachusetts: Harvard University Press. Available at: http://www.hup.harvard.edu/catalog.php?isbn=9780674725867 (Accessed: 13 November 2018).

Wang, T. (2013). Big Data Needs Thick Data, Ethnography Matters. Available at: http://ethnographymatters.net/blog/2013/05/13/big-data-needs-thick-data/ (Accessed: 13 November 2018).

Womack, J. P., Jones, D. T. & Roos, D. (1990) *The Machine That Changed the World*. Simon and Schuster. Available at: https://www.lean.org/Bookstore/ProductDetails.cfm?SelectedProductID=160 (Accessed: 13 November 2018).

Chapter 4
Designing a design conference in an emerging design science community: Danish experiences from the International Conference on Design (DESIGN)

Tim C. McAloone and Mogens Myrup Andreasen

Abstract The International Conference on Design, "DESIGN" is an important reoccurring event in the lives and careers of many design researchers and practitioners, worldwide. The conference has acted as a catalyst for many conversations, discussions, presentations and working meetings about the nature, contents and intent of design, as a vital activity in the continuous development of the synthetic world. This article provides a balance on the DESIGN conference and its contribution to the research community that today is gathered within the Design Society. We compare and contrast DESIGN against the Design Society's flagship conference, ICED, and other conferences and meetings affiliated with the Design Society. And we reflect on the way in which the DESIGN conference has acted as a conduit and a testbed for the design research discussions and results that have emerged from our own university, the Technical University of Denmark.

4.1 The establishment of DESIGN

The DESIGN conference pre-dates the consolidation and establishment of the Design Society, which was officially founded in 2000 – and first announced at DESIGN 2000. The DESIGN conference was among the activities and groups that constituted the initial bouquet of regular reasons for members to meet in the newly established Design Society. Operating biennially on even-numbered years and alternating every year with the Design Society's own flagship conference, the International Conference on Engineering Design (ICED), DESIGN has become a regular conference for Design Society members and other colleagues from the design research community to gather, share ideas and results, network and plan collaboration activities. Although viewed

T. C. McAloone (✉) · M. M. Andreasen
Technical University of Denmark, Kongens Lyngby, Denmark
e-mail: tmca@dtu.dk

D. Marjanović et al. (eds.), *Design Research: The Sociotechnical Aspects of Quality, Creativity, and Innovation*, https://doi.org/10.1007/978-3-031-50488-4_4

by many as a Design Society event, the unique characteristic about DESIGN is that it is wholly owned, managed and run by a relatively small group of academics at the University of Zagreb's CADLab. Since its inception in 1981, the conference has gained in size, reputation, and global representation, within the design research community.

From the very beginning, the collaboration with the community now known as the Design Society (and before that the WDK: Workshop Design-Konstruktion) has been very close. Already from the second event in 1984, a strong collaboration between Zagreb University and WDK was formed, after WDK's Vladimir Hubka presented as invited speaker and dialogue began about collaboration.

4.2 Continuity as a key characteristic

With its unique position as a conference that is owned and organised by the same team each time the conference is organised, a number of interesting learnings can be gained, regarding the quality of the organisation, quality of the papers and the general experience for the delegate. Compared to the ICED conference series, which is coordinated by the Design Society's incumbent Board of Management and 'franchised' each time to willing bidders from amongst the Design Society's membership, DESIGN has a permanent ownership, a permanent organising committee and a continuing programme committee, from conference to conference. The charm and attraction of ICED is that the delegate experiences a different location and a different local flavour for each new conference edition, with new forms of communication and dissemination being experimented, and a large emphasis on the local organisers' chosen specific theme of the conference. In contrast, the attraction of DESIGN is the comfort of having a known programme structure, the known location and the relatively stable selection of activities and communication forms. Having much more experience in the organisation and running of the practical contents of DESIGN leaves room for continuous improvements and innovations within the programme, which result, when looking back over past three decades since the conference began, to a unique set of conference experiences for the delegate. In the following, we reflect on some of these unique and highly professional contents of the DESIGN conference.

4.3 What makes DESIGN unique?

DESIGN has led the way within the Design Society, regarding setting the standard with respect to ***article reviewing***. Year-on-year improvements to both article review forms and the actual instructions and feedback to both reviewers and authors has led to a gradual and sustained increase in the quality of the reviews provided to contributors. A constant pruning of the reviewer pool, plus a careful matching of articles to reviewers' research areas and fields of interest has led to an extremely high hit-rate, regarding both amount of papers reviewed (both on-time and at all) and the quality of the reviews themselves. Recent years have seen the addition of the functionality within the paper review system, for authors to provide feedback to reviewers and enter into a dialogue about the quality of the paper and the review process; this practice is highly acclaimed by authors to the DESIGN conference. All of the above tweaks and efforts to improve the reviewing system are paying off; the quality of submitted papers (measured as a function of the rejection rate) increases year-by-year. The latest DESIGN conference (2024) saw the highest quality papers at the DESIGN conference yet, with an acceptance rate of 59% (of all papers submitted). Both Scopus and Web of Science (two important indexing agencies, according to many universities' performance indicators) have indexed DESIGN's papers since 2012, which is no mean feat for a conference series to achieve.

The DESIGN conference was the first within the Design Society to fully integrate ***workshops*** into the conference programme, attracting dedicated participants to half-day, in-depth exchanges, discussions and exercises, often organised by Design Society's Special Interest Groups (SIGs), but also open for others to try and test an idea or spark off the interest for a future SIG. The style of each workshop is left to the individual workshop leader, giving an exciting platform for lectures, discussions, tutorials, sketching-sessions and paper-writing activities. Workshops are always planned for the first day of the conference, giving delegates the chance to attend two half-day workshops, and boosting the communication and knowledge sharing, from the very start of the conference.

Another tradition that has emerged at DESIGN, becoming a fixed part of the conference inventory is the ***PhD forum***, where currently active PhD students get the opportunity to share their experiences, concerns and ideas, all helped by the coaching of senior academic delegates from the conference—professors whom the PhD students respect and look up to. Informal round-table presentations and focus-discussions build the PhD students' networks and boost their confidence, reassuring them that they're not alone in their experiences and that there's normally a way to work through what to them may seem as a very new and unknown problem or dilemma.

The conference's ***coffee breaks and lunch sessions*** are not to be underestimated, not because of the quality of the coffee (which frankly isn't the best in Europe!), but due to the length of time given to network among the delegates. Long coffee breaks of 30 minutes and extended lunches of 75 or 90 minutes are not a waste of time—they allow for longer dialogue between delegates, deeper reflection on the session that just ensued and the chance to catch up with colleagues and meet new friends. They also give the slightly nervous young researcher the chance to approach the speaker from a previous session and pose the question they might not have had the courage to do so in plenum (Fig. 4.1)

Whilst other conference series waver between whether to award best paper awards or not, DESIGN has consistently recognised ***reviewers' favourites*** for the past 12 years. These have been awarded to the conference's top 5% or 10% of papers, according to the score awarded by reviewers, in the double-blind review process.

Fig. 4.1 SIG workshops and PhD forum

Such awards are important for young researchers and established professors alike—many a framed certificate has been identified, hanging on the walls of research offices, around the world!

Perhaps one of the most unique and relatively new traditions at DESIGN is the ***DESIGN Debate***; a staged debate where a motion is posed and two members of the Design Society's community support the motion, where two others oppose the motion. The topic is always contentious (e.g. "The motion is that design researchers should stop wasting time and important resources by making useless research experiments on students, instead of industry", 2018), the debaters prepare themselves excruciatingly thoroughly, and the audience has great fun, whilst at the same time getting the chance to reflect on both sides of a design research related topic (Fig. 4.2).

The sense of community within DESIGN is a factor that is extremely important to its Conference Chair. For this very reason, the ***conference size*** has always been a number to keep under control. If Conference Chair does not know everyone or has at least had the chance to meet everyone by the end of the conference, then there were too many delegates! As a nice contrast to its sister conference, ICED, where average delegate numbers are reaching 700 and parallel sessions can amount to ten, DESIGN is intentionally about half this size in both dimensions, allowing for the discussions, sessions and social activities to be organised to allow for maximum interaction and deep, informal dialogue.

Fig. 4.2 Casting votes at the DESIGN Debate!

4.4 The changing nature of a community over time

Across the Design Society in general, an observation over recent years has been that the topics submitted and discussed within our community have changed almost unrecognisably over the past four decades. When the ICED and DESIGN conferences began, much of the research focus was on machine elements, mathematical design modelling and computer aided design considerations. To illustrate this, the five volumes of the Proceedings of ICED 1990 largely contained papers on machine elements and mechanical engineering; by ICED 1999 such papers occupied just one of the volumes.

Today's conferences include a majority of higher-level, design management, and diversely represented fields such as human factors, product/service-systems, product architecture, decision-making, industrial design. The 'traditional' contents is still to be found in DESIGN, but in fewer numbers today, and sometimes under the guise of robust design or design optimisation. The Zagreb team behind DESIGN have provided moving images of word clouds and affinity matrices over the past decade or so, allowing us to see how the subjects emerge, rise and then fall, with respect to delegates' research interests. There is no certainty as to whether the Design Society (which constitutes a high percentage of DESIGN's delegates) has lost focus, is fashion-driven regarding societal fads and trends, or whether the ever-powerful journals have taken over the role of the deep discipline related subjects such as finite elements and machine elements, leaving space for the broader discussion of interdisciplinary design during the conferences.

Another noteworthy observation is the gradual emergence of the topic of 'design research' within the conference itself. Differentiating itself from 'research in the field of engineering design', 'design research' places its focus on understanding, through research, the actual process of designing. The Design Society-affiliated 'Summer School on Engineering Design', which has a number of PhD students attending both the summer school and the DESIGN conference, introduced the topic 'How to do design research?' in the mid 1990s. And in 2001, Blessing and Chakrabarti published their seminal book, Design Research Methodology (Blessing & Chakrabarti, 2009), providing a strong structure to the structuring and support of design research. The theme of design research has permeated DESIGN ever since, if not as a sole article topic in itself, then as a structure for the methodology in a large number of the papers published in the conference.

The contents of the DESIGN conference are of course a direct reflection of the delegates and their professional backgrounds. As design has grown as a field of research and practice over the years, so too has its network. From its initial audience some 28 years ago of primarily engineering design researchers and practitioners, today the DESIGN conference (and the Design Society alike) has followers and members from industrial design, anthropology, applied psychology, systems engineering and design management, to name but a few. The increasing diversity of delegates and their papers, most of whom seem to see the proposed themes for a DESIGN conference as inspiration rather than a framework for the conference, leads to an increasing need to articulate what we mean by the word and the practice of 'design'. How far can we and should we allow the field to stretch, before the collective potential to fruitfully share across disciplines no longer holds as feasible? A question we must constantly seek the answer to, year-on-year.

One unique characteristic of the DESIGN conference has perhaps been its ability, through the conference's Programme Committee, to deal with the 'delta' between the call for papers at the very genesis of a given conference edition, all the way to the final themes and sessions of a DESIGN conference. The flexibility of the Programme Committee to observe patterns of research, as they emerge through the submitted papers and to cluster and organise these into sessions, is an important asset of the DESIGN conference and a key activity of the conference design and development. Again, the advantages of a stable Programme Committee, year after year, are manifest, through the group's knowledge of each other and their experience from previous conference years.

4.5 How has one research school—DTU—navigated DESIGN, over time?

First, some history: Fifty years ago, in 1968, Dr. Vladimir Hubka joined a small group of five faculty (Jeppesen, Tjalve, Stahl, Boe and Andreasen) at the Technical University of Denmark (DTU), where he worked for two years, before moving on to Switzerland. When Hubka joined DTU, the main focus at the time was on establishing design courses and consolidating a comprehensive design methodology. One could argue that Hubka's time at DTU marked the beginning of DTU's design research activity; he brought a mature theory and methodology on engineering design to the table. When Hubka moved in 1970 to ETH in Zürich, the collaboration broadened, with Andreasen (DTU) collaborating with Hubka (ETH) and Umberto Pighini (University of Rome). Through this collaboration, WDK was established as the framework for cooperation around research into engineering design. The collaboration under WDK eventually leading to the first of many ICED conferences, hosted in Rome 1981, by Pighini.

First from the 1980ies, DTU began to educate PhD candidates in the area of engineering design and product development, thus introducing the question of how to do design research. This activity of design research steadily grew over the decades, with a careful eye for how to combine design research, design teaching and design practice.

DTU's current situation: Today, the design research effort at DTU boasts a whole of three design research groups, with a total of 3 professors, 17 associate or assistant professors, around 20 post-doctoral or other contract-employed researchers and over 30 PhD students. The 'Copenhagen School', as it is occasionally referred to, thus consists of a number of disciplines, all of which have design research as their main research identifier—and all of whom in some way or other interface with the university's students, as part of the teaching effort towards Mechanical Engineering Masters Design and Innovation Engineering Masters, or Industrial Engineering and Management Masters.

DTU has in some way or other always been connected to the Design Society (and WDK, before it), and to the DESIGN conference. Thus, it is interesting to trace the development over time, of DTU's contributions to the DESIGN conference and observe how one representative organisation has developed over the years, with respect to its contribution and research development.

DTU's contribution and development to DESIGN: The following is an 'archaeological exercise' through the past 16 years of DESIGN proceedings, where the number of articles and the topics covered within the articles have been charted. A short discussion accompanies this review exercise, of topics as they have evolved. The method used to carry out this review involved an 'author and affiliation' identification, where any contribution that included an author from DTU (either as main author or co-author) was included. It should be noted that DTU employees from all three above-mentioned research groups are counted. The topics charted were aggregated according to the current DESIGN conference themes and topics, so as to allow for a consistent charting, back in time.

Figure 4.3 depicts DTU's contributions to the DESIGN conference in the years 2002-2018. The left-hand column shows research topics, found by reading the articles submitted and matching these to keywords that have arisen throughout the 16-year period under examination. The subsequent columns show where papers have covered one of the research topics (marked with a dot per article), in a particular conference year. As one paper often covers more than one research topic, the total number of dots for a year always exceeds the total number of papers submitted by DTU for that year, summarised on the third-to-last row of the figure. The bottom rows show how many of the published papers were submitted by DTU (or rather, with a DTU author affiliated), how many papers were published in that particular year, and the percentage of papers submitted by DTU, for a given conference year. Finally, the right-hand column sums the occurrence of dots for a given research topic (showing frequency and therefore popularity of a topic) and the bottom-right figures show the overall submission statistics, over the 16 years.

4.5.1 Analysis of DTU's contributions to DESIGN

In the following, we make a short analysis of the above review of DTU contributions to DESIGN, 2002–2018. The small pictograms, below, depict the area of focus for the discussion and should simply act as a guide to studying Fig. 4.3.

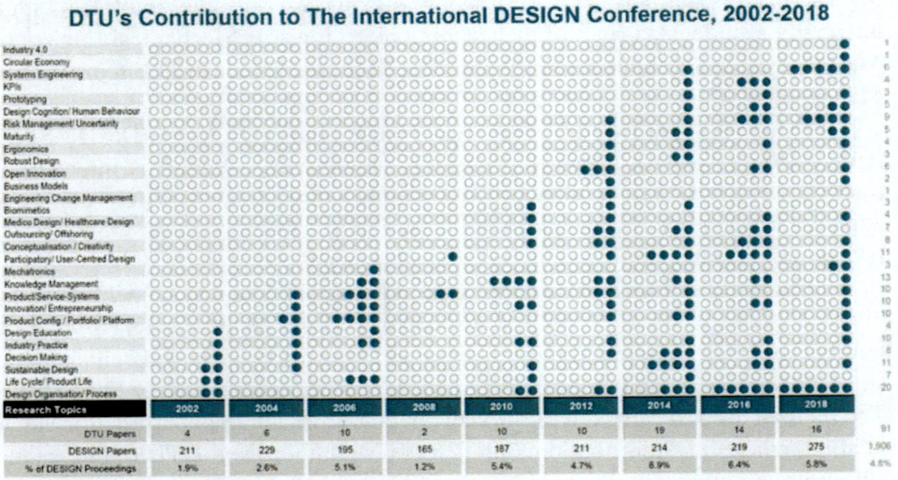

Fig. 4.3 DTU contributions to DESIGN, years 2002–2018

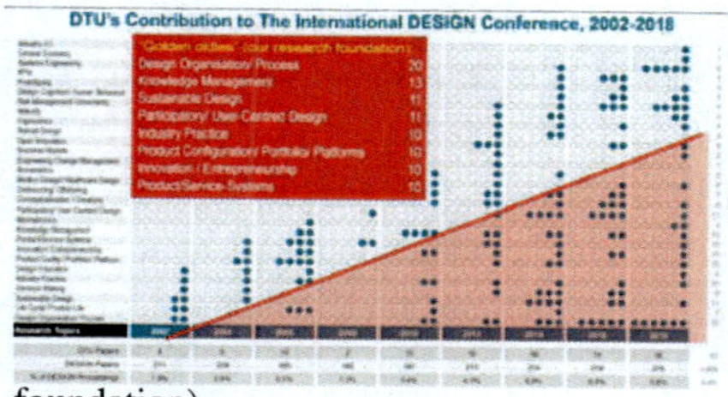 (our research foundation)	**The "Golden oldies"** Observing the bottom-right triangle of DTU's contributions to DESIGN, it becomes apparent, which research topics are the foundation of the DTU authors, with respect to recurrent themes. Interestingly, the most recurrent theme is Design Organisation/ Design Process (20 papers); followed by Knowledge Management (13); Sustainable Design (11); Participatory/User-Centred Design (11); Industry Practice (10); Product Configuration/ Portfolio/Platforms (10); Innovation/Entrepreneurship (10); Product/Service-Systems (10). The "real classic" WDK/Design Society themes of engineering design and machine elements are not the most occurrent over this period; these were more popular prior to 2002 and are gaining slight increase in recent years (as Robust Design).
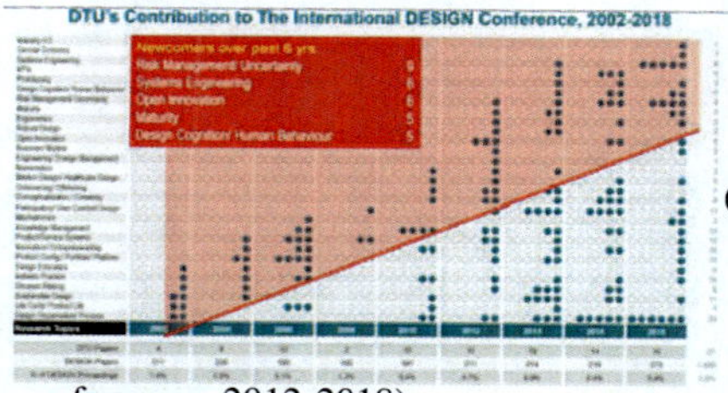 (over past four conferences, 2012-2018).	**Newcomers** The past six years has seen a significant increase in five thematic areas, in particular, which did not exist in DTU's design research portfolio earlier. These include Risk Management/ Uncertainty (9 papers); Systems Engineering (6); Open Innovation (6); Sustainability Maturity (5); Design Cognition/Human Behaviour (5). These additional research topics show a tendency towards higher-level research themes, more trans-disciplinary research and more societally contributory research topics.

(continued)

(continued)

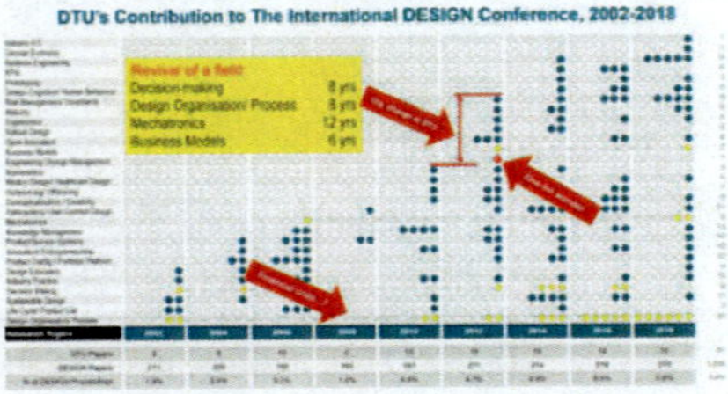	**"One-hit wonders", a short gap, plus revival of some themes** If one should look carefully at the pictogram to the left, one will be able to see a red dot. This red dot shows one paper at one DESIGN conference, on the topic of Engineering Change Management, which has not been in focus before or after the 2012 conference. This is DTU's one-hit-wonder, at least in the context of the DESIGN conference. In 2008, there were just two papers from DTU at DESIGN, marking an abnormally low amount of contributions. Finally, four research themes (marked with yellow dots in the pictogram, to the left) returned to the DESIGN conference, after significant gaps. These themes are Mechatronics (gap of 12 yrs.); Decision-Making (8 yrs.); Design Organisation/ Design Process (8 yrs.); Business Models (6 yrs.). Interestingly, an organsiational change at DTU, resulting in an increase of design research related staff in 2011-12, can be seen in the 'spike' of new research topics in 2012 (from the red dot, upwards).
	**Design for societal impact** DTU has always had a large proportion of design research topics that relate to themes of direct societal impact (as opposed to e.g. basic or experimental laboratory research). These topics, coloured in orange, to the left include: Participatory/User-Centred Design (11 papers); Sustainable Design (11); Innovation/ Entrepreneurship (10); Industry Practice (10); Design Education (4); Ergonomics (4); Medico Design/Healthcare Design (4); Business Models (2); Industry 4.0 (1); Circular Economy (1). Total count: 58

(continued)

(continued)

	**Measuring and modelling performance** Slicing the review in a slightly different way, it can be seen that a number of DTU's research contributions to DESIGN focus on measuring and modelling the performance of design and product development (green lines in the pictogram). Papers related to these topics include: Knowledge Management (13 papers); Industry Practice (10); Design Cognition/Human Behaviour (5); Maturity (5); KPIs (4); Robust Design (3); Engineering Change Management (1). Total count: 41
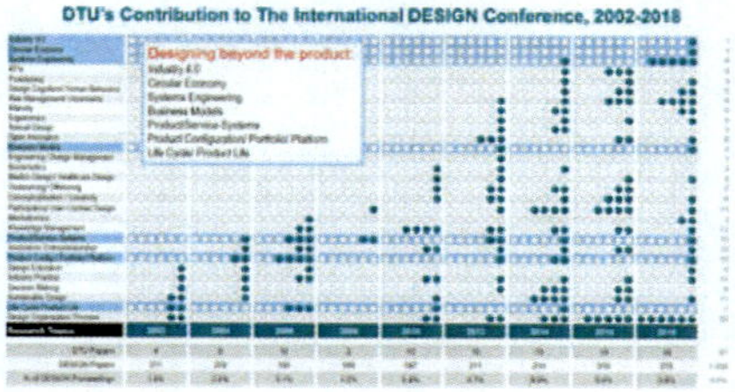	**Designing beyond the product** Yet another view on DTU's research can be seen when looking beyond 'product design' to other more augmented or high-level design objects in our research (blue lines in pictogram). These include: Product/Service-Systems (10 papers); Product Configuration/Portfolio/Platform (10); Life Cycle/Product Life (7); Systems Engineering (6); Business Models (2); Industry 4.0 (1); Circular Economy (1) Total count: 37
	**Good design practice** The final slice of the review focuses on DTU's research attention to Design Practice (brown lines in pictogram), which is one of the DESIGN conference's recurring themes. Here, DTU's research contributions include: Design Organisation/Process (20 papers); Risk Management/Uncertainty (9); Conceptualisation/Creativity (8); Decision Making (8); Outsourcing/Offshoring (7); Open Innovation (6); Biomimetics (3); Mechatronics (3); Prototyping (3). Total count: 67

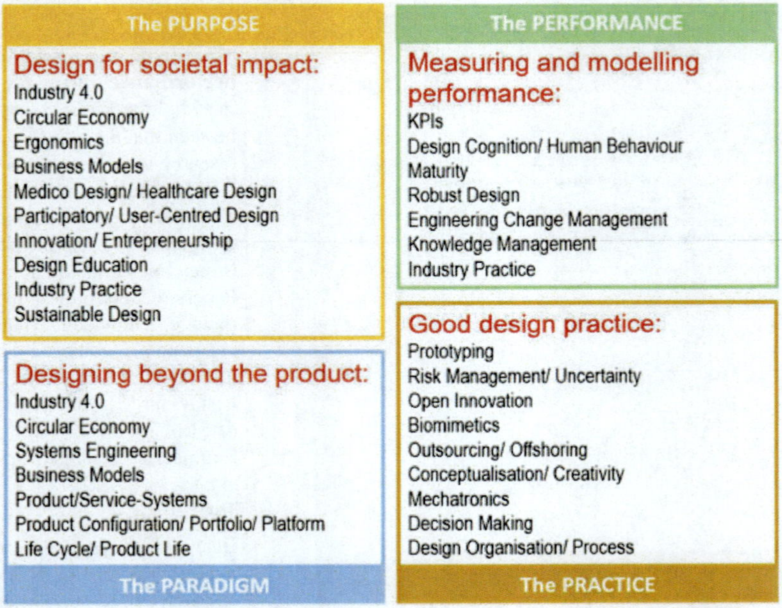

Fig. 4.4 DTU's Contribution to The International DESIGN Conference, 2002-2018

4.5.2 Discussion

The DESIGN conference has undoubtedly played a significant role in the development of DTU's design research agenda, acting as an arena to develop, test and present our research outputs and create a dialogue with peers. At the same time, we can also claim that DTU has had an impact on the DESIGN conference over time, with 91 papers listed in the references (or 4.8% of the contributions in the period 2002-2018) being authored or co-authored by a DTU colleague.

From the analysis of the contributions carried out, above, one can see a general trend of DTU's design research, over the years. And with a little goodwill, one can categorise DTU's design research into four main areas, of: (i) the PURPOSE of design; (ii) the PERFORMANCE of design; (iii) the PRACTICE of design; and (iv) the PARADIGM of design, as depicted in Fig. 4.4

4.5.3 What did we see from the contributions?

It is interesting to see from the analysis, that DTU has a broad contribution to design research topics, and is influencing the contents and thematic focus of the DESIGN conference. The areas where DTU has grown over the years correspond in many cases, to the areas where DESIGN has also developed. The areas where DTU has its strongest focus, at least at the DESIGN conference, are within Design PRACTICE and Design PURPOSE. The areas of Design PERFORMANCE and Design PARADIGM are also areas where DTU contributes to the DESIGN conference, albeit to a lesser extent.

4.5.4 What did we NOT see from the contributions?

Firstly, it is important to remember that the analysis of the design research results carried out in this paper applies to just a portion of the research output from the same group of colleagues, and not ALL of their research output. Some research topics (e.g. robust design, circular economy, product/service-systems, design education) have comprehensive 'competing' outlets to DESIGN, in the form of dedicated conferences to the respective topics. To gain a totally balanced overview of the same colleagues' research outputs over the same period, therefore, would require a comprehensive review exercise, broader than their DESIGN contributions. Equally and oppositely, it is not easy to conclude that DTU no longer contributes as heavily to e.g. engineering design and machine elements, from the analysis of the DESIGN papers alone, although this tendency is actually true.

Secondly, the 'publish or perish' phenomenon has firmly taken hold over the past decade, significantly changing the motivation for researchers to publish in journals vs. conferences, and resulting in a total transition at DTU, to article-based PhD theses, instead of monographies. At DTU, we have worked closely to maintain a healthy balance between publishing to conferences and journals, ensuring our PhD students of the quality and invaluable feedback one can achieve from one's peers, by submitting research to conferences, such as DESIGN, rather than merely publishing to journals. But it is a challenge that is not showing any signs of subsiding, any time soon, and the types (scope, breadth, depth, maturity) of contributions from DTU to DESIGN have changed in nature, over the past decade.

Thirdly, as with all such analyses, one cannot see the effect of colleagues coming and going, from the results. Certain areas with a strong representation for a shorter period (e.g. open innovation, medico-/healthcare design) followed by termination, correspond to a lead researcher's relatively short period of employment at DTU, whereas others (e.g. design process, sustainable design) reflect a longer-career DTU'er.

A fourth area that one does not find in the above analysis is the existence and deeply motivating contents of the invaluable workshops that the DESIGN conference invites and hosts. These topics, largely reflecting the Design Society's SIGs, are not at all reflected in the paper analysis carried out here.

Finally, the analysis does nothing to reflect the many personal collaborations, new project ideas, good discussions over a cup of coffee, or long-lasting friendships that occur during DESIGN, making this event one of the most significant community-building venues within the design research community.

Fig. 4.5 DTU's Contribution to Design research, going forward? Some possible trajectories

4.5.5 What might we see in the coming years?

Who dares to look into the future and project a research direction? We do! DTU actually practices a rolling research strategy and planning activity, where we, year-by-year, ask ourselves which direction we think our research within the university is heading. Building on our basis (bottom-left of Fig. 4.5 we feel it safe to say that we will focus on the contribution of design and design research to the development of the world, in the form of understanding and responding to mega-trends and industrial/societal affinities. Most of all, we can guarantee that our research will be focusing on how the most important mega-trend of them all—*the sustainability of our society and the planet*—can be positively influenced and supported through design.

4.6 Final reflections

On reflection of this balancing paper, where we have highlighted the DESIGN conference's key characteristics, its merits and its contributions to our community, we believe that the DESIGN conference has had—and continues to have—a clearly significant role in the development of the community that we identify as the Design Society. Among the Design Society's seven objectives, the DESIGN conference has a direct contribution to five of these, namely: to create and evolve a formal body of knowledge regarding design (through the fantastic back-catalogue of conference results); to support and improve design research and education; to promote cooperation among design researchers, practitioners, educators and managers; to promote design publications and their dissemination; and to organise international conferences and other events. DESIGN is truly a conference series that has an important and lasting contribution to the Design Society.

We reflect that the smaller, more informal, close and familiar style of DESIGN stands as a beautiful complement to the ICED conference series, which offers almost opposite conditions for many characteristics. For this reason too, the DESIGN conference gives a similar, yet different platform to discuss and develop research ideas and results. The difference between ICED and DESIGN does not lie in a topic- or thematically-oriented differentiation, but in an organisational and staging differentiation.

The flexibility of DESIGN (and of the Design Society), over time, has meant that design research as a field has been able to develop and grow, supported by the conference as an outlet for research ideas and results, but not restrictive in regards to keeping last year's structure as a strait jacket for contributions at subsequent events. The organisation of the conference is *absolutely* professional and a clear example of world-class organisation and leadership—elegantly mixing formalised routine, continuous improvement and best practice with the human touches of flexibility, opportunism, collegial spirit and sense of humour.

Our own research groups at DTU have reaped fantastic benefit from the DESIGN conference over the years, using the event as a platform to share, discuss, recruit, collaborate and contribute through many different media, provided by the conference.

Both authors of this paper have been honoured to sit on the Programme Committee of DESIGN, where we have experienced how the backstage organisation of this conference runs like clockwork—and where we have experienced the balance of routine and opportunism, all the way down to the way in which papers are prepared, sorted, reviewed and subsequently processed, by a team that seems from the outside to be much, much larger than it actually is. We therefore join in congratulating 'Zagreb Team' for their outstanding contribution to the design research community and for a world-class conference series, by the simple name of DESIGN.

References

Achiche, S., Howard, T. J., Ástvaldsdóttir, A., Andersen, O., & McAloone, T. C. (2012). Open innovation practices in a cluster context: A Medicon valley case study. In *Proceedings of 12th International Design Conference—Design 2012*.

Achiche, S., Howard, T. J., & McAloone, T. C. (2012). The underlying relationships between success criteria & success factors in NPD activities. In *Proceedings of 12th International Design Conference—Design 2012*, pp. 1–10.

Ahmed, S., & Hansen, C. T. (2002). An analysis of design decision-making in industrial practice. In *Proceedings of 7th International Design Conference—Design 2002*, pp. 145–150.

Ahmed, S. (2006). Prompting designers to design. In *Proceedings of 9th International Design Conference—Design 2006*.

Ahmed-Kristensen, S., & Christensen, B. T. (2014). Naturally original: stimulating creative design through biological analogies and random images. In *Proceedings of 13th International Design Conference—Design 2014*, pp. 427–436.

Ahmed-Kristensen, S., & Soendergaard, E. (2016). Developing a support tool for global product development decisions. In *Proceedings of the DESIGN 2016 14th International Design Conference* (Vol. DS 84, pp. 727–738).

Andreasen, M. M., Wognum, N., & McAloone, T. C. (2002). Design typology and design organisation. In *Proceedings of 7th International Design Conference—Design 2002.*

Andersen, J. B., Howard, T. J., & McAloone, T. C. (2014). Development of a data capture tool for researching tech entrepreneurship. In *Proceedings of 13th International Design Conference—Design 2014*, pp. 1–10.

Andersen, S. N. (2016). The process of participatory ergonomics simulation in hospital work system design. In *Proceedings of the DESIGN 2016 14th International Design Conference.*

Bay Brix Nielsen, C. K. E., Cash, P., & Daalhuizen, J. (2018). The behavioural design solution space: examining the distribution of ideas generated by expert behavioural designers. In: Marjanović, D., Štorga, M., Škec, S., Bojčetić, N., & Pavković, N. (eds.), *Proceedings of the DESIGN 2018 15th International Design Conference*, pp. 2739–2750).

Bey, N. (2002). Environmental assessment—gotten across to industrial designers. In *Proceedings of 7th International Design Conference—Design 2002*, pp. 1293–1298.

Blessing, L.T.M., & Chakrabarti, A. (2009). *DRM, a Design Research Methodology*. Springer-Verlag.

Bruun, H. P. L. (2014). Mapping requirements to a product Architecture supported by a PLM system. In *Proceedings of 13th International Design Conference—Design 2014*, pp. 1741–1752.

Cash, P., & Kreye, M. E. (2014). Information processing theory in the early design stages. In *Proceedings of 13th International Design Conference—Design 2014*, pp. 455–464.

Chong, K. N., Dreijer, L. S., Howard, T. J., Birkved, M., Kreye, M. E., Bey, N., & McAloone, T. C. (2014). Comparing LCC With LCA to Assess PSS Sustainability: The Case of the Eco-Box. In *Proceedings of 13th International Design Conference—Design 2014*, pp. 783–790.

Christensen, M. E., Howard, T. J., & Rasmussen, J. J. (2012). The foundation for robust design: Enabling robustness through kinematic design and design clarity. In *Proceedings of 12th International Design Conference—Design 2012.*

Conceicao, C., Lundsgaard, C., & Broberg, O. (2014). Developing through prototyping: A resource material on user involvement for workspace design. In *Proceedings of 13th International Design Conference—Design 2014.*

Cramer-Petersen, C. L. (2016). Argumentation and reasoning in design: an empirical analysis of the effects of verbal reasoning on idea value in group idea generation. In *Proceedings of the DESIGN 2016 14th International Design Conference* (Vol. DS 84, pp. 857–966).

Edwards, K. (2004). Expected and realized costs and benefits when implementing product configuration systems. In *Proceedings of 8th International Design Conference—Design 2004*, pp. 216–231.

Ernstsen, S. K., Thuesen, C., Larsen, L. R., & Maier A. (2018). Identifying disruptive technologies: Horizon scanning in the early stages of design. In D. Marjanović, M. Štorga, S. Škec, N. Bojčetić, & N. Pavković (Eds.), *Proceedings of the DESIGN 2018 15th International Design Conference*, pp. 2739–2750.

Fan, Z., Matzen, D., Andreasen M. M., & Hein, L. (2006). Teaching system integration of mechatronic systems. In *Proceedings of 9th International Design Conference—Design 2006*, pp. 1133–1142.

Fiil-Nielsen, O., & Mortensen, N. H. (2006). Implementing product platforms: A case study. In *Proceedings of 9th International Design Conference—Design 2006*, pp. 823–830.

Hansen, C. T., Andreasen, M. M. (2004). A mapping of design decision-making. In *Proceedings of 8th International Design Conference—Design 2004* pp. 1409–1418.

Hansen, C. T., & Andreasen, M. M. (2010). On the content and nature of design objects in designing. In *Proceedings of 11th International Design Conference—Design 2010*, pp. 761–770.

Hansen, Z. N. L., & Ahmed-Kristensen, S. (2010). The impact on the product development process when offshoring or outsourcing. In *Proceedings of 11th International Design Conference—Design 2010.*

Hansen, Z. N. L, & Ahmed-Kristensen, S. (2012). Connecting global product development with corporate strategy. In *Proceedings of 12th International Design Conference—Design 2012*, pp. 465–474.

Hansen, C. L., Mortensen, N. H., & Hvam, L. (2012). On the market aspect of product program design: Towards a definition of an architecture of the market. In *Proceedings of 12th International Design Conference—Design 2012.*

Hein, L. (2004). The role of design in university induced industrial innovation. In *Proceedings of 8th International Design Conference—Design 2004* (pp. 267–272).

Howard, T. J., Achiche, S., Özkil, A., & McAloone, T. C. (2012). Open design and crowdsourcing: Maturity, methodology and business models. In *Proceedings of 12th International Design Conference—Design 2012*, pp. 181–190.

Jensen, A. R. V., & Ahmed-Kristensen, S. (2010). Identifying knowledge in decision-making processes: a case study. In *Proceedings of 11th International Design Conference—Design 2010*, pp. 1543–1552.

Jensen, T. E., & Andreasen, M. M. (2010). Design Methods in Practice—Beyond the 'Systematic Approach' of Pahl & Beitz. In *Proceedings of 11th International Design Conference—Design 2010.*

Jensen, C. S., Jensen, A. R., & Broberg, O. (2016a). Intervention framework to support employee-driven innovation between R&D and Manufacturing department. In *Proceedings of the DESIGN 2016 14th International Design Conference* (Vol. DS 84, pp. 1055–1064).

Jensen, L. S., Özkil, A. G., & Mortensen, N. H. (2016b). Prototypes in engineering design: definitions and strategies. In *Proceedings of the DESIGN 2016 14th International Design Conference* (Vol. DS 84, pp. 821–830).

Jensen, L. S., Nissen, L., Bilde, N., Ozkil, A. G. (2018). Prototyping in Mechatronic Product Development: How Prototype Fidelity Levels affect User Design Input. In: Marjanović, D., Štorga, M., Škec, S., Bojčetić, N., & Pavković, N. (eds.), *Proceedings of the DESIGN 2018 15th International Design Conference*, pp. 2739–2750.

Kadir, B. A., Broberg, O., & Souza da Conceicao, C. (2018). Designing Human-Robot Collaborations in Industry 4.0: Explorative Case Studies. In: Marjanović, D., Štorga, M., Škec, S., Bojčetić, N., & Pavković, N. (eds.), *Proceedings of the DESIGN 2018 15th International Design Conference*, pp. 2739–2750.

Kjeldal-Jensen, O., & Ahmed-Kristensen, S. (2010). Informing early-phase technology decisions in paradigmatic innovation. In *Proceedings of 11th International Design Conference—Design 2010*, pp. 1553–1564.

Kunrath, K., Cash, P., & Li-Ying, J. (2016). Designer's identity: personal attributes and design skills. In *Proceedings of the DESIGN 2016 14th International Design Conference* (Vol. DS 84, pp. 1729–1740).

Kunrath, K., Cash, P., Li-Ying, J. (2018). Designers' identity: skills' self-perception and expectation in design students. In: Marjanović, D., Štorga, M., Škec, S., Bojčetić, N., & Pavković, N. (eds.), *Proceedings of the DESIGN 2018 15th International Design Conference*, pp. 2739–2750.

Langer, S., Maier, A. M., Wilberg, J., & Münch, T. J., Lindemann, U. (2012). Exploring differences between average and critical engineering changes: survey results from Denmark. In *Proceedings of 12th International Design Conference—Design 2012.*

Larsson, F., Mortensen, N. H., & Andreasen, M. M. (2004). Recognizing the needs for improving the portfolio management for new products in the industry. In *Proceedings of 8th International Design Conference—Design 2004*, pp. 337–342.

Larsson, F. (2006). Declining productivity in product development: An explanation of "pipeline gridlock". In *Proceedings of 9th International Design Conference—Design 2006*, pp. 1533–1542.

Lasso, S. V., Cash, P., Daalhuizen, J., & Kreye, M. (2016). A model of designing as the intersection between uncertainty perception, information processing, and coevolution. In *Proceedings of the DESIGN 2016 14th International Design Conference* (Vol. DS 84, pp. 301–310).

Lenau, T. A., Dentel, A., Ingvarsdóttir, Þ., & Guđlaugsson, T. (2010). Engineering design of an adaptive leg prosthesis Using biological principles. In *Proceedings of 11th International Design Conference—Design 2010.*

Li, X., & Ahmed-Kristensen, S. (2014). Product development in China: Comparison between Danish and Chinese companies. In *Proceedings of 13th International Design Conference—Design 2014*, pp. 1581–1590.

Lokkegaard, M., Mortensen, N. H., Jensen, L. S., & Christensen, C. K. F. (2018). Assessing increased product line commonality's effect on assembly productivity and product quality. In:Marjanović, D., Štorga, M., Škec, S., Bojčetić, N., & Pavković, N. (eds.), *Proceedings of the DESIGN 2018 15th International Design Conference* pp. 2739–2750.

Lund Strom, L. C., Willumsen, P. L., Oehmen, J., & Heck, J. (2018). Can design thinking mitigate critical strategy implementation risks? In: Marjanović, D., Štorga, M., Škec, S., Bojčetić, N., & Pavković, N. (eds.), *Proceedings of the DESIGN 2018 15th International Design Conference*, pp. 2739–2750.

Marini, V. K., & Restrepo, J. (2010). Evaluation of information requirements of reliability methods in engineering design. In *Proceedings of 11th International Design Conference—Design 2010*, pp. 845–854.

Marini, V. K., & Ahmed-Kristensen, S. (2012). Decision-making and feedback as foci for knowledge-based strategies supporting concept development. In *Proceedings of 12th International Design Conference—Design 2012.*

Mata, M. P., Ahmed-Kristensen, S., & Brockhoff, P. B. (2014). influence of consumer's background on product perception. In *Proceedings of 13th International Design Conference—Design 2014.*

Matzen, D., & Andreasen, M. M. (2006). Opportunity parameters in the development of product/ service-systems. In *Proceedings of 9th International Design Conference—Design 2006*, pp. 929–936.

McAloone, T. C. (2002). Creating sustainable students through project-based teaching. In *Proceedings of 7th International Design Conference—Design 2002.*

McAloone, T. C., & Andreasen, M. M. (2004). Design for utility, sustainability and societal virtues: developing product/service-systems. In *Proceedings of 8th International Design Conference—Design 2004*, pp. 1545–1552.

McAloone, T. C., Hansen, P. H .K., & Larsen, J. H. (2004). Images of innovation—an ontological approach. In *Proceedings of 8th International Design Conference—Design 2004*, pp. 53–60.

McAloone, T. C., Mougaard, K., & Restrepo, J., Knudsen, S. (2010). Eco-innovation in the value chain. In *Proceedings of 11th International Design Conference—Design 2010*, pp. 855–864.

Mougaard, K., Howard, T. J., McAloone, T. C., Neugebauer, L., & Bey, N. (2012). Establishing collaborative networks for the conceptualisation of PSS. In *Proceedings of 12th International Design Conference—Design 2012*, pp. 249–258.

Munk, L., & Mortensen, N. H. (2006). Platform evaluation using explicit modelling: A case study. In *Proceedings of 9th International Design Conference—Design 2006* (pp. 951–958).

O'Hare, J., & McAloone, T. C. (2014). Eco-innovation: the opportunities for engineering design research. In *Proceedings of 13th International Design Conference—Design 2014.*

Okholm, A. B., Rask, M., Christensen, M. E., Eifler, T., Holmbeg, M., & Howard, T. J. (2014). Improving process capability database usage for robust design engineering by generalising measurement data. In *Proceedings of 13th International Design Conference—Design 2014.*

Omidvarnia, F., Hansen, H. N., Bissacco, G., & Islam, A. (2014). Integrated design as an approach towards DFμMA. In *Proceedings of 13th International Design Conference—Design 2014.*

Parraguez Ruiz, P., Eppinger, S. D., & Maier, A. (2014). Evolution of information control and centralisation through stages of complex engineering design projects. In *Proceedings of 13th International Design Conference—Design 2014*, pp. 1329–1338.

Parraguez Ruiz, P., & Maier, A. (2016). Network insights for partner selection in inter-organisational new product development projects. In *Proceedings of the DESIGN 2016 14th International Design Conference.*

Pedersen, R., & Mortensen, N. H. (2006). Possible Explanations to the limited use of product platforms in industry. In *Proceedings of 9th International Design Conference—Design 2006.*

Piccolo, S. A., Wilberg, J., Lindemann, U., & Maier, A. (2018). Changes and sentiment: A longitudinal email analysis of a large design project. In: Marjanović, D., Štorga, M., Škec, S., Bojčetić, N., & Pavković, N. (eds.), *Proceedings of the DESIGN 2018 15th International Design Conference*, pp. 2739–2750.

Pieroni, M., Pigosso, D. C. A., & McAloone, T. C. (2018). Exploring the synergistic relationships of circular business model development and product Design. In Marjanović, D., Štorga, M., Škec, S., Bojčetić, N., & Pavković, N. (eds.), *Proceedings of the DESIGN 2018 15th International Design Conference*, pp. 2739–2750.

Pigosso, D. C. A., & McAloone, T. C. (2014). Systematization of best practices for Ecodesign implementation. In *Proceedings of 13th International Design Conference—Design 2014*, pp. 1651–1662.

Pigosso, D. C. A., Pattis, A. T., McAloone, T. C., & Rozenfeld, H. (2014). Deployment and implementation of the grundfos sustainability strategy by means of the ecodesign maturity model. In *Proceedings of 13th International Design Conference—Design 2014*, pp. 1663–1670.

Pikas, E., Oehmen, J., Koskela, L., & Thuesen, C. (2018). A new framework for construction project definition stage. In: Marjanović, D., Štorga, M., Škec, S., Bojčetić, N., & Pavković, N. (eds.), *Proceedings of the DESIGN 2018 15th International Design Conference*, pp. 2739–2750.

Ravn, P. M., , Gudlaugsson, T. V., & Mortensen, N. H. (2014). Visual modelling of pilot production to support decision making in production development. In *Proceedings of 13th International Design Conference—Design 2014.*

Restrepo, J. (2006). Assessing relevance: designers' perception of information usefulness. In *Proceedings of 9th International Design Conference—Design 2006*, pp. 1235–1242.

Restrepo, J., McAloone T. C., Schlegel, T., Lykke, J. (2008). A user-centered approach to developing emergent technology products. In *Proceedings of 10th International Design Conference—Design 2008.*

Rodrigues, V. P., Morioka, S., Pigosso, D. C. A., Carvalho, M., & McAloone, T. C. (2016a). Exploring the dynamic and complex integration of sustainability performance measurement into product development. In *Proceedings of the DESIGN 2016 14th International Design Conference* (Vol. DS 84, pp. 433–442).

Rodrigues, V. P., Pigosso, D. C. A., & McAloone, T. C. (2016b). Process-oriented performance indicators for measuring ecodesign management Practices. In *Proceedings of the DESIGN 2016 14th International Design Conference* (Vol. DS 84, pp. 443–452).

Rodrigues, V. P., Pigosso, D., & McAloone, T. C. (2018). Linking Ecodesign capabilities to corporate performance: proposal of a simulation-based Approach. In: Marjanović, D., Štorga, M., Škec, S., Bojčetić, N., & Pavković, N. (eds.), *Proceedings of the DESIGN 2018 15th International Design Conference*, pp. 2739–2750.

Sarka, P., Ipsen, C., Maier, A. M., Heisig, P., & Caldwell, N. H. (2014). Engineers are using social media for work purposes. In *Proceedings of 13th International Design Conference—Design 2014.*

Soendergaard, E. S., & Ahmed-Kristensen, S. (2014). Decision making in global product development. In Proceedings of 13th International Design Conference—Design 2014.

Stavrakos, S. K., & Ahmed-Kristensen, S. (2012). Assessment of anthropometric methods in headset design. In *Proceedings of 12th International Design Conference—Design 2012.*

Stavrakos, K., Ahmed-Kristensen, S., & Goldman, T. (2014). Using archetypes to create user panels for usability studies. In Proceedings of 13th International Design Conference—Design 2014, pp. 2157–2166.

Sudin, M. N., & Ahmed-Kristensen, S. (2010). The role of a specification in the design process: A case study. In *Proceedings of 11th International Design Conference—Design 2010* , pp. 955–964.

Tan, A. R., & McAloone, T. C. (2006a). Understanding and developing innovative products and services: The essential elements. In *Proceedings of 9th International Design Conference—Design 2006*.

Tan, A. R., & McAloone, T. C. (2006b). Characteristics of strategies in product/service-system development. In *Proceedings of 9th International Design Conference—Design 2006*.

Tan, A., Andreasen, M. M., & Matzen, D. (2008). Conceptualisation of product/service-systems through structural characteristics. In *Proceedings of 10th International Design Conference—Design 2008*.

Taylor, T., & Ahmed-Kristensen, S. (2014). The applicability and coherence of key performance indicators in global product development. In *Proceedings of 13th International Design Conference—Design 2014*.

Taylor, T., & Ahmed-Kristensen, S. (2016). Global product development: KPI selection support. In *Proceedings of the DESIGN 2016 14th International Design Conference* (Vol. DS 84, pp. 1615–1624).

Tegeltija, M., Oehmen, J., Kozine, I., & Geraldi, J. (2016). Post-probabilistic uncertainty quantification: discussion of potential use in product development risk management. In *Proceedings of the DESIGN 2016 14th International Design Conference* (Vol. DS 84, pp. 533–542).

Tegeltija, M., Oehmen, J., McMahon, C. A., Maier, A., Kozin, I., Škec, S. (2018). Tailoring risk management in design. In: Marjanović, D., Štorga, M., Škec, S., Bojčetić, N., & Pavković, N. (eds.), *Proceedings of the DESIGN 2018 15th International Design Conference*, pp. 2739–2750.

Thorpe, J. R., Forchhammer, B. H., & Maier, A. (2016). Needs elicitation for novel pervasive healthcare technology. In: *Proceedings of the DESIGN 2016 14th International Design Conference* (Vol. DS 84, pp. 1947–1956).

Valentin-Hjorth, J. F., Patou, F., Syhler, N., Dominguez, H., & Maier, A. (2018). Design for health: Towards collaborative care. In Marjanović, D., Štorga, M., Škec, S., Bojčetić, N., & Pavković, N. (eds.), *Proceedings of the DESIGN 2018 15th International Design Conference* (pp. 2739–2750).

Valgeirsdottir, D., Onarheim, B., & Li-Ying, J. (2016). improving creativity training: a study of designer skills. In: *Proceedings of the DESIGN 2016 14th International Design Conference* (Vol. DS 84, pp. 1175–1184).

Vianello, G., Xie, Y., Ahmed-Kristensen, S., & Culley, S. J. (2010). Handling of in-service support: Comparison of two case studies from complex industries. In *Proceedings of 11th International Design Conference—Design 2010*.

Wied, M., Oehmen, J., & Welo, T. (2018). Resilient Design Properties of a Driverless Transport System. In: Marjanović, D., Štorga, M., Škec, S., Bojčetić, N., & Pavković, N. (eds.), *Proceedings of the DESIGN 2018 15th International Design Conference* , pp. 2739–2750.

Willumsen, P. L., Oehmen, J., & Ernstsen, S. K. (2018). Reconceptualizing Design Risk Management as a Learning Strategy. In D. Marjanović, M. Štorga, S. Škec, N. Bojčetić, & N. Pavković (eds.), *Proceedings of the DESIGN 2018 15th International Design Conference* (pp. 2739–2750).

Witz, P., & Oehmen, J. (2018). Making partnerships work: Integrative open system design for a new generation of complex infrastructure schemes. In: Marjanović, D., Štorga, M., Škec, S., Bojčetić, N., & Pavković, N. (eds.), *Proceedings of the DESIGN 2018 15th International Design Conference*, pp. 2739–2750.

Chapter 5
Dynamics of Using Information and Communication Technology Tools in a Distributed Project-Based Design Course

Nikola Horvat, Niccolò Becattini, Harshika Singh, and Stanko Škec

Abstract This chapter presents the dynamics and interplay of individual and collaborative use of ICT tools during a design project-based learning course. During the COVID-19 pandemic, five teams (8–9 members) with members from Croatia, Italy, Slovenia, and Austria worked in a geographically distributed context. The results show that students utilise both individual and collaborative working modes. However, the dynamics of ICT use differ throughout the course. Typical design tools for 3D modelling become predominant just in the later course phases. In contrast, visualisation tools are used more during the initial phases. Finally, findings suggest that remote collaboration requires continuous and frequent use of communication tools (video conferencing and instant messaging) throughout the whole course. The presented findings can help educators and researchers in engineering education better structure their classes in post-pandemic settings.

5.1 Introduction

The recent changes that the COVID-19 pandemic has brought to our lives highlighted several weaknesses of our society. At the same time, the pandemic also highlighted the importance of being prompt to ideate and develop solutions to problems that abruptly pop up and bring a revolution in everybody's routine. In this reference, nobody can get out of those troubles independently and, despite the requirements of social distancing, the need of collaborating to generate strategies and solutions that have an impact became paramount. However, students have few opportunities to collaborate during their study course and most of their collaboration skills are built

N. Horvat (✉) · S. Škec
Faculty of Mechanical Engineering and Naval Architecture, University of Zagreb, Zagreb, Croatia
e-mail: nikola.horvat@fsb.hr

N. Becattini · H. Singh
Department of Mechanical Engineering, Politecnico Di Milano, Milan, Italy

D. Marjanović et al. (eds.), *Design Research: The Sociotechnical Aspects of Quality, Creativity, and Innovation*, https://doi.org/10.1007/978-3-031-50488-4_5

up outside of the standard courses, typically due to students' association initiatives, which are mostly conceived for socializing, rather than for building up skills and capacities.

In this perspective, the Project-Based Learning (PBL) approach (Dym et al. 2005) offers to the educators the opportunity to let students improve their skills and competencies through a learning-by-doing pedagogy that puts them in front of real (or realistic) project tasks that are often also complex and ill-structured by their own definition. Through this kind of pedagogic approach students work with peers (typically with the same background). They address a problem for which they need to find a solution and so they can learn how to work in a team, how to bring their own competencies to the team, and how to collaborate to achieve different objectives that converge towards a unique project goal. Moreover, they can experience and practice how to interact live and remotely, how to plan and manage the work to be done as well as how to familiarize with the professional tools they will use once they access their position in the job market.

Still on the side of the changes introduced by the COVID-19 pandemic, this health-related social revolution clearly pushed the academic institution and the higher education system at large to rethink their teaching process completely. What was typically carried out with traditional ex-cathedra lectures, with a direct interaction between the educator and the learner, now gets mediated by the technology and by a wide variety of ICT systems, as these were immediately perceived as the most direct and effective way to reach out geographically distributed classes of students and prevent the spread of the virus. The pandemic, therefore, pushed the need for computer-mediated communication to the extremes and at the same time offered a huge opportunity to leverage the use of ICT tools for remote collaboration. In other words, this created the conditions to shed light on the interplay between the collaborative learning approach typical of PBL (Verstegen et al. 2016) with remote learning, as all the activities needed to be carried out by means of information and communication technology (ICT) tools due to the social restrictions in place to contain the spread of the virus. The exclusive use of ICT tools to carry out every task in the project, without any other form of live interaction allowed, is a unique chance to clarify what are the tools, e.g., by the functionalities they offer, that facilitate or impair the inherent activities that students have to carry out within the project. Moreover, the wide variety of tasks that the students have to carry out during a design project makes them need functionalities that are typically offered by different ICT tools to achieve their project goal. Nevertheless, the proficient use of these ICT tools might also depend on the familiarity that students already have with them (Brisco et al. 2016), as the steepness of these learning curves can trigger frustration and limit their adoption (Brisco et al. 2020a, 2020b; Verstegen et al. 2016). This means that different students, or teams of them, can react very differently to the need of using a variety of ICT tools during the execution of a design course structured according to the PBL pedagogy (Horvat et al. 2021). However, the previous studies limited their focus on adopting the use of tools according to the needs of the different stages of the process with a cross-sectional perspective, neglecting the progression of the usage of these ICT tools during the project. The analysis of how the use of ICT tools changes during the design process, capturing the behaviour of engineering designers with a more longitudinal perspective, despite within a not excessively long time-frame, is necessary in order to highlight the strengths and weaknesses of these systems, beyond the role they play in the different project stages.

Therefore, this contribution offers an original investigation on the usage of ICT tools, as it aims at unveiling similarities and differences among individuals and among design teams during the execution of PBL course organized around a design project, in order to highlight dynamics and peculiarities with reference to the different functionalities these tools offer. The related findings are relevant to educators/researchers in engineering education, professionals, and developers of ICT tools for designing. The first ones can benefit from insights enabling more structured support to students during design PBL courses. The second ones can identify potential pitfalls due to the execution of collaborative design projects in remote settings. The third ones, eventually, can exploit the findings to develop ICT tools that have higher chances to match the requests of the market dealing with geographically distributed design teams.

The next section provides additional details on the two relevant sides of this contribution: the application of the PBL pedagogy in engineering design courses as well as the use of ICT tools to support these practices. The third section describes the method proposed to assess the degree of interaction that different teams of students had with ICT tools during the execution of a geographically distributed PBL design course. Section four presents the results of the investigation with two complementary perspectives: the general trend of students participating in the course and the different behaviour that characterizes the five project teams during the execution of their project activities. Before concluding the paper, these results will be discussed in order to depict some general implications that can be relevant both for engineering design educators to improve their pedagogical approaches and for professionals that aim at making their (geographically) distributed project activities more effective and efficient.

5.2 Background

5.2.1 Project-Based Learning (PBL) Courses in Engineering Design

Project-based learning (PBL) courses are organised around the main project assignment or central problem, which organises and drives students towards a tangible outcome and product. These project assignments and central problems serve as a starting point of the learning/teaching process for students/educators. For that reason, they need to be carefully considered and formulated to provide students with the opportunity to obtain a wider set of learning outcomes. Usually, as one of the main characteristics of the PBL approach (Kolmos and de Graaff, 2007), these assignments are complex or ill-defined, originating from industry (Becattini et al. 2020; Vukašinović and Pavković, 2017) or students themselves.

As one of the most utilised educational approaches in an engineering design domain, PBL offers several educational benefits such as an increase of student engagement and participation, communication skills, critical thinking, industrial and real-life relevance of student assignment (Grimheden and Hanson, 2005). Often, these student-centred learning-by-doing courses include working in smaller teams to collaboratively try to explore and understand the problem, analyse the existing and propose new solutions and, finally, evaluate and refine the proposed solution (Kolmos and de Graaff, 2007). Therefore, PBL facilitates collaborative learning and provide means to introduce a social-constructivist paradigm in a learning process (Verstegen et al. 2016). For that reason, it is of utmost importance for PBL courses to establish working structures and provide an environment in which students with different backgrounds and previous knowledge can collaborate (Lou and Kim MacGregor, 2004), both in an online or traditional classroom environment. Based on Brisco et al. (2019), two different aspects of PBL courses need to be emphasised. First, reflection represents an important part of design education and practice (Reymen et al. 2006), and students should be able to elaborate, discuss and edit the design being developed with other team members. Second, students should be able to become familiar with different design topics individually (self-learning) and share responsibilities while working on the same design within a team context. The understanding of this interplay between individual and team activities within design PBL courses remains an unaddressed challenge.

These various characteristics and advantages of the PBL approach are in accordance with the design education tendencies. Therefore, many educators within the design education community were willing to introduce it to their courses. For example, courses such as EGPR (Vukašinović and Pavković, 2017), GPD (Leung et al. 2019), the Global Design Project (Brisco et al. 2019) or ELPID (Becattini et al. 2020; Horvat et al., 2021) allow students to work on a project with an industrial partner in a geographically distributed environment. These courses follow the traditional product development process—planning, concept development, embodiment, detailing and prototyping, and include several milestones (after each phase). Also, they foster collaborative work in teams and put emphasis on different aspects of human-centred design and creative problem-solving. Besides being multidisciplinary, these courses are conducted in a distributed manner and require additional organisational effort due to various challenges which might cause collaboration and communication issues. These course settings are, to a certain extent, similar to the needs and challenges of engineering educators emphasised during the COVID-19 pandemic: familiarity of educators with online tools, addressing student preferences, assessment procedures, supporting e-learning infrastructure for students (and educators), issues with hands-on and project-based education, etc. (Asgari et al. 2021). However, the COVID-19 pandemic required a sudden transition from face-to-face education to completely online delivery of courses within a short time frame. These COVID-19 transitions were often conducted ad-hoc and the majority of educators experienced various difficulties, while the abovementioned pre-pandemic distributed courses were systematically developed and delivered for a certain number of years. Therefore, the analysis of these distributed courses, which were initiated pre-pandemic, could offer complementary insights and indicate potential ways of improvement for suddenly transitioned courses, especially in terms of addressing collaboration and communication challenges.

To help students overcome these collaboration barriers and resolve underlying issues (such as e.g., insufficient motivation, language barriers, online communication issues), in their recent study, Brisco et al. (2019) explored the way students could be supported with different collaboration guidelines. Their workshop findings indicate the importance of simple communication methods, clear team protocols, synchronous work, project management and knowledge management techniques, etc. Using a retrospective survey, Mamo et al. (2015) explored how students used different design and communication tools throughout the course and provided some findings on the distribution of individual and collaborative work in design courses. However, longitudinal studies on ICT usage should be performed to gain new and detailed insights in terms of the dynamics of individual and teamwork conducted by students.

Within the virtual environment, reliance on the usage of ICT tools and recent ICT advancements in terms of data collection and monitoring, provides educational researchers with new opportunities to identify and explore different aspects of individual and collaborative work throughout virtual design PBL courses.

5.2.2 *Use of ICT Tools in a PBL Courses in Engineering Design*

Within the context of virtual design PBL courses, various ICT tools have been used for general communication, specialised design work (e.g., Computer-Aided Design (CAD), Computer-Aided Engineering (CAE)), project management and file sharing. Moreover, in a geographically distributed environment, and consequently virtual, students are forced to collaborate with different ICT tools, and they do not possess any other means to fulfil course objectives. For that reason, these tools have a massive role in the design of PBL courses, which was further emphasised by the COVID-19 pandemic. In addition, familiarity with these design tools is perceived as one of the most important skills students need to acquire during their engineering education. For all these reasons, the usage of ICT tools in design PBL courses attracted a lot of attention in the recent period.

Related studies are mostly focused on analysing a series of courses to examine general ICT practices (e.g., Brisco et al., (2020a, 2020b)) or exploring the usage of specific ICT tools within design PBL courses (e.g., Becattini et al., (2020)). Brisco et al. (2016) explored the extent to which students use various communication platforms for PBL courses and the underlying rationale for their selection. Their findings indicate inappropriate usage of cloud file repositories, incompatibility issues between different tools and influence of previous tool familiarity on its selection. Within the same line of argumentation, Verstegen et al. (2016) emphasised potential issues with student motivation and interest in case learning a new ICT tool is perceived as too demanding (information overload). In a related study, by asking students, Brisco et al. (2019) identified collaboration challenges, current issues with the functionalities of available technologies and future recommendations. This study points out that some identified challenges cannot be resolved solely by technology, but that misuse and inadequate selection of technology can significantly impede the course project progress. Studies related to the usage of specific ICT tools are mostly oriented towards exploring their functionality and applicability for course purposes. For example, Albers et al. (2009) explored the general usage of wikis, PDM, and CAD without analysing their utilisation dynamics. On the other hand, by considering PBL and e-learning principles, Becattini et al. (2020) proposed a complete e-learning infrastructure to deliver PBL courses. Later, Horvat et al. (2021) reflected on the use of design ICT tools and e-learning platform elements during the virtual course.

In the PBL courses, the majority of deliverables are created as digital objects (e.g., CAD models, project reports, presentations and videos), and to generate them, students have to utilise ICT tools. As such, monitoring of using digital tools on a more granular level could provide detailed insights and a better understanding of the role that different ICT tools have in the design process performed by students. Previous educational studies indicate that there are different approaches on how data collection can be performed related to the usage of ICT tools—tool use analysis (Gopsill et al. 2017), retrospective interviews/questionnaires (Mamo et al. 2015) or case studies (Vukašinović et al. 2011). However, to obtain longitudinal study information, there is a necessity to conduct a series of data collection instances to identify potential changes or trends over a certain period. As such, longitudinal studies can offer several benefits such as real-time analysis, sequence identification, repeated observations, and elimination of recall bias (Robinson, 2012). Since previous studies have been mainly oriented towards analysis on the whole course level or analysing specific collaboration aspects within the course, the main objective of this study is to better investigate the underlying dynamics of ICT usage over the course period.

5.3 Methods

The presented work is designed as a longitudinal study with work sampling as the data collection procedure (Robson and McCartan, 2016). The study analyses the use of the six most commonly used ICTs during a semester-long PBL design course. The studied ICTs are video conferencing (e.g., MS Teams, Zoom), instant messaging (e.g., MS Teams chat, WhatsApp), CAD (e.g., Onshape, SolidWorks), file repository

Table 5.1 Studied variables and their measures

ICT tools category	Description	Scale
Video conference	Using video conference tools such as MS Teams or Skype	1–7 frequency Likert
Instant messaging	Using instant messaging tools	1–7 frequency Likert
CAD tools—Individual	Individual work on CAD tools	1–7 frequency Likert
CAD tools—Collaborative	Collaborative work on CAD tools	1–7 frequency Likert
File repository—Individual	Using file repository tools such as OneDrive	1–7 frequency Likert
Visualisation—Individual	Individual use of visualisation tools such as collaborative whiteboard (e.g., Miro)	1–7 frequency Likert
Visualisation – Collaborative		1–7 frequency Likert
Task management – Individual	Using task management tools such as Trello	1–7 frequency Likert
Task management – Collaborative	Using task management tools such as Trello	1–7 frequency Likert

(e.g., OneDrive, Google Drive), visualisation (e.g., Miro, Mural) and task management (e.g., Trello, Asana). These ICTs are often found in the PBL design courses, and a recent e-learning framework has been proposed around them (Becattini et al., 2020). Their usage is measured with a self-reporting questionnaire that builds on 7 points frequency Likert scale (1—Never, 2—Rarely (less than 10%), 3—Occasionally (about 30%), 4—Sometimes (about 50%), 5—Frequently (about 70%), 6—Usually (about 90%), 7—Always). Table 5.1 shows all the studied variables with their description and scale.

5.3.1 Course and Participants Description

The use of tools has been studied during an internationally distributed PBL design course. The course is organised as a collaboration among four universities (Politecnico di Milano, TU Wien, University of Ljubljana and University of Zagreb) in which five virtual student teams work on the design problem introduced by a partner company (Siemens Mobility). Four student teams were comprised of eight members, while one team had nine members. The members within each team were evenly distributed among the four universities (i.e., two members from each university) except for nine-membered team B that had three members from one university and two from the other three universities. In total, four female and 37 male mechanical engineering students on both undergraduate and graduate levels participated in the course. Each team had one or two academic coaches who worked as the team's facilitators during the course. Two academic coaches were also researchers that analysed this data, being, in a way, ethnographic researchers within their respective two teams.

The course is organised as a competition amongst five virtual teams who had to design new seats and seating arrangements in metros following the design course phases (Table 5.2). Siemens Mobility introduced this course problem during an initial six-day workshop. Besides problem introduction, the workshop served to help students form teams, transfer knowledge about early design phases (user and market research, brainstorming), and familiarise them with the problem context. The partner company delivered several lectures to explain the main aspects of the metro design (e.g., general overview, safety, acoustics, fire protection). These lectures enabled students to reach a contextual understanding and consider more requirements while designing. Furthermore, design-related lectures (user and market research, brainstorming) were organised in the form of active participation, as students had to work by applying introduced methods to their task. To help them work, a course staff proposed a tool for each ICT category introduced by Becattini et al. (2020), namely Microsoft (MS) Teams for video conferencing, instant messaging and file repository, Miro for visualisation, while the CAD tool varied among teams. Hence, for the CAD tool, Team A used Onshape with some initial attempts with other CAD tools, Team B and C used SolidWorks, Team D used 3DExperience and Autodesk Fusion 360, and Team E used Onshape and Siemens NX.

After the initial online workshop, the course followed three phases (identification of opportunities, conceptual design, embodiment design) separated by formal review meetings involving company representatives. For each of these review meetings, students prepared a presentation and a report. During the first phase, students

Table 5.2 ELPID course structure and timeline

Phase name	Scope	Deliverable	Duration
Initial workshop	Introduction of problem, context, design methods and ICT tools. Initial market research	Chosen market in the form of presentation	1.5 weeks
Identification of opportunities	User and market research, overview of existing solutions. Proposing few visions and requirements for the solution	Few visions and requirements for the solution in the form of report and presentation	3 weeks
Conceptual design	List of requirements, problem clarification, functional decomposition, concept generation, concept evaluation	Few concepts and requirements for the solution in the form of report and presentation	4 weeks
Embodiment design	CAD modelling, strength calculations, rendering	Proposed solution in the form of report, presentation and CAD models	5 weeks
Final presentation preparation	Pitch presentation preparation	Final pitch in the form of presentation	2 weeks

conducted user and market research, patent screening, and comparison of an existing solution to propose few visions on how they would like to proceed. They also developed an initial list of requirements that define the problem space. In the second phase, students developed several concepts for a chosen vision. After the conceptual design phase, teams worked on the embodiment design, after which they had an additional deliverable in the form of a virtual prototype for their solution. Finally, two weeks after the third review, the teams presented their final design to the broader company staff. In the final presentation, the teams incorporated rendered images and a video of their solution.

5.3.2 Data Collection

The data on the usage of tools was collected with a questionnaire through seven sampling periods (SPs) distributed across the course (see Fig. 5.1). Each sampling period lasted for two course weeks, starting at the course week 3. Given that each phase was roughly 4 weeks in duration, this sampling density enabled analysis on the use of ICT tools within a course phase, as well comparison between phases with the reduced load on the respondents as compared to more fine-grained analysis. Hence, every two weeks, the team coaches distributed the electronic questionnaire to their team. As the first two course weeks were reserved for the initial workshop, which differed in its structure from the rest of the course, the data has not been collected for this period. To increase the response rate, few days after the sampling point, the coaches reminded students who did not fill in the questionnaire to do so. The questionnaire was shared using MS Forms and was designed such that it takes around 10 min to complete. The measured variables with the questionnaire as well as their scale are provided in Table 5.1.

In total, 255 responses were collected across seven sampling periods, distributed as follows: SP1–42 responses, SP2–35 responses, SP3–37 responses, SP4–36 responses, SP5–37 responses, SP6–41 responses, SP7–27 responses. The distribution of these responses within each sampling period, together with mapping to course

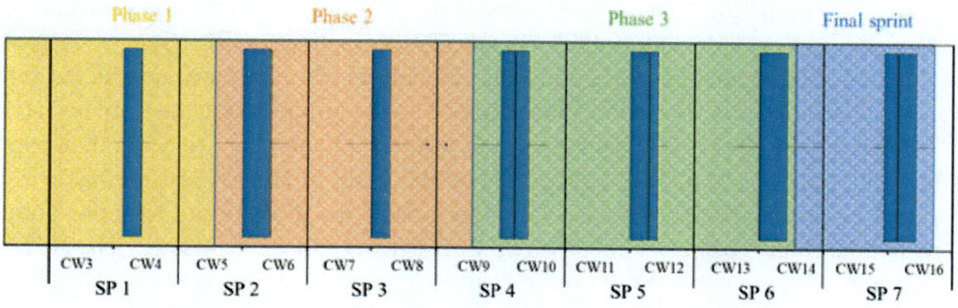

Fig. 5.1 Response distribution across sampling periods (SPs), course weeks (CWs) and course phases

weeks and phases, are provided in Fig. 5.1. Response distribution in each sampling period is presented with a horizontal boxplot. In the boxplot, the rectangles represent responses between the lower (Q1) and upper (Q3) quartile, while the whiskers represent the earlier and later responses within the sampling period. In addition, a vertical line represents the median value of the responses (note that for the SP1 and SP2 median is equal to the upper quartile, while for the SP3 and SP6 to the lower quartile).

5.3.3 Data Processing Procedure

Given that the data was collected from a questionnaire that was supposed to be filled by a sample of 41 students for seven sampling points, there could be some inconsistencies in the data. The first issue was that the students did not simultaneously fill in the questionnaire after its distribution. Consequently, a span of several days corresponds to one sampling point. Therefore, the sampling periods are introduced and defined as 14-day periods with a mean date corresponding to the sampling point. Each period started seven days before the sampling point and lasted for seven days after the sampling point. Second, a data cleaning procedure was developed to deal with the multiple answers of a team member within one sampling period, thus preventing skewing the results. Duplicate answers for each participant within each sampling period were identified and then averaged to get the mean value of the responses.

After the data had been cleaned, two verification steps were conducted on a subset of the data. The first verification step included qualitative matching of the participants' responses to the researchers' experience while coaching the team. This verification step was possible due to the researchers being incorporated as coaches of two teams. Their experience in coaching a team enabled an understanding of the patterns related to the dynamics of the ICT tools usage and the interplay between individual and collaborative work. The second verification step included mapping the responses to objective measures of the ICT tools usage (if available). This step was available for teams that used Onshape as a CAD modelling tool since it estimates the modelling time. The modelling time on the team level could be mapped to the self-reporting data and thus provide additional verification of the self-reporting data.

After cleaning and verifying the data, an analysis procedure was conducted at the course level (i.e., all respondents were treated as one group) and at the team level (i.e., data was analysed for each team separately). As this chapter aims to better understand the longitudinal use of ICT tools, the data has been processed separately for each sampling period. Hence, for each sampling period, the average and variation among respondents have been calculated on the course level and the team level. The analysis thus includes a comparison of average values in seven sampling periods for each ICT mentioned above. This comparison of averages will provide an initial estimation of the longitudinal use of ICT tools during the PBL course, while variations will provide a better understanding of the individual ICT use within an analysed group.

5.4 Results

This section presents students' usage of ICT tools on the course level (Sect. 5.4.1) and the team level (Sect. 5.4.2) throughout the product design PBL course. The analysis at the course level will show dynamic patterns of the ICT tools usage that emerge when all the teams are combined. On the other hand, the more in-depth analysis at the team level will result in similarities and differences between teams in using tools during the PBL design course. Furthermore, separate analyses for individual and collaborative work in both subsections will show the interplay between these two working modes.

5.4.1 Use of ICT Tools on the Course Level

Students' individual use of ICTs varied across categories. For the individual use, task management, CAD, file repository and visualisation tools usage data has been collected (Fig. 5.2). For example, the use of the task management tools slightly decreased as the course advanced, with the highest use at the beginning of the course. On the other hand, the individual use of the CAD tools barely existed at the beginning, but their use increased in the later course phases, with the peak in the course weeks 13 and 14. Students used file repositories throughout the course phases. However, their usage varied the most among these four tools, suggesting that portion of the course participants used it, while others did not. Finally, individual use of visualisation tools (e.g., collaborative whiteboard) was persistent until the course week 10, after which started to decline.

Besides individual use, students used tools collaboratively in order to work together (Fig. 5.3). For instance, communication tools (video conferencing and instant messaging) were crucial for their continuous information exchange as they were used persistently throughout the course. Video conferencing tools were used the most frequently and consistently. As the course proceeded, their use slightly decreased. Nevertheless, they still stayed the most used tools category at the end of the course. On the other hand, instant messaging was used slightly more in the later phases of the course, but still less than video conferencing. Similar to the individual

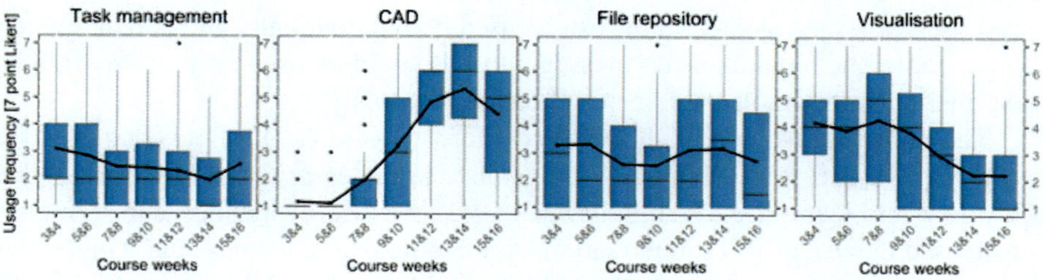

Fig. 5.2 Individual use of ICT tools

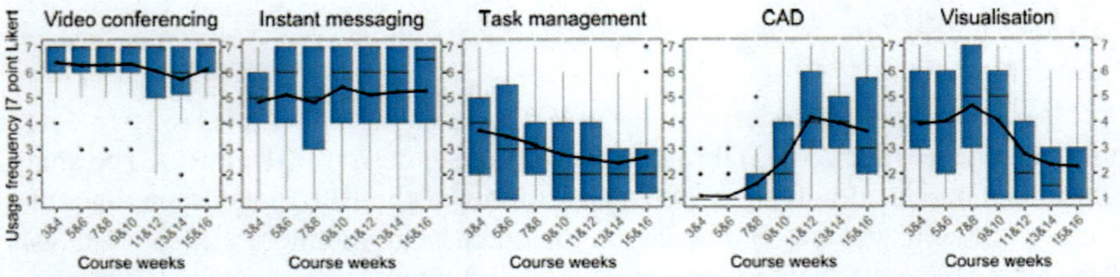

Fig. 5.3 Collaborative use of ICT tools

use of task management tools, their collaborative use also decreased as the course proceeded. The collaborative use of the CAD tools was non-existent in the first six course weeks, then started to increase until course weeks 11 and 12, after which it gradually declined. Collaborative visualisation tools were used more often until the course weeks 9 and 10 (peak at course weeks 7 and 8), and then started to decline.

The dynamics of the individual and collaborative use does not differ much. However, differences exist in the extent of the collaborative usage of tools as compared to individual use. For example, even though the dynamics of the CAD and visualisation tools are similar for both usage types, course participants used CAD slightly more individually than collaboratively. On the other hand, it seems that the visualisation tools are used slightly more collaboratively than individually. Furthermore, while the peak positions are aligned for the individual and collaborative visualisation tools usage (weeks 7 and 8), the peak of the collaborative CAD use was slightly earlier (weeks 11 and 12) than the peak of the individual CAD use (weeks 13 and 14). Hence, tools on the course level were used both individually and collaboratively, with some of them supporting all the weeks, while others were supporting only specific weeks. However, the use of task management and visualisation tools had high variation, and more specific analysis on the team level is necessary to unravel whether these differences are within or between teams.

5.4.2 *Use of ICT Tools on the Team Level*

To identify differences as well as similarities between teams, the analysis on the team level for each ICT tools category has been conducted. While visualisation, CAD and task management tools can be used both individually and collaboratively, video conferencing and instant messaging tools could be used only collaboratively, and file repository tools could be used only individually.

The collaborative use of the video conferencing tools (Fig. 5.4) suggests that teams A, B and E started higher with their use, while teams C and D started slightly lower. While team A only slightly decreased their use over time, teams B and E decreased the use of ICTs much more. On the other hand, teams C and D increased their use of video conferencing over time. Furthermore, while teams A, B and C had a gradual change in the usage of video conferencing tools, teams D and E had periodical change every two weeks. However, these periodical changes in the use of video conferencing were not consistent for the two teams. On the one hand, team D had lower use of video conferencing in weeks 3 and 4 and higher in weeks 5 and 6. On the other hand, team E's use was inversed, i.e., higher in weeks 3 and 4 and lower in weeks 5 and 6. Furthermore, teams A and D had lower variation in the usage of video conferencing tools, while teams B, C and E had higher variation. Given these similarities and differences, it can be concluded that all the teams extensively used video conferencing tools but at a slightly different frequency. Further research is necessary to unravel the reasons behind these differences between teams.

The instant messaging tools (Fig. 5.5) were not used as frequently as video conferencing. However, their use was also consistent throughout the course. Teams A and D increased the use, while others slightly decreased the usage of instant messaging tools (teams C and E) as the course proceeded. Compared to video conferencing, the use of instant messaging was not as uniform since the variations within a team are usually high. These results suggest that while all the members consistently used video conferencing tools, some members used instant messaging tools more than others. Finally, it is worth mentioning that team E was the only one that decreased their use of both communication modes throughout the semester, suggesting the lower communication as the course advanced.

For most teams (except team E), the individual use of task management tools gradually decreased over time (Fig. 5.6). The exception was team E, who gradually increased the use of task management tools during the last six weeks of the course. However, since variation within team E was high on these sampling points, it might be that only a subset of members started using the task management tools. Furthermore, the most frequent use of these tools was reported by the team C. Team D reported

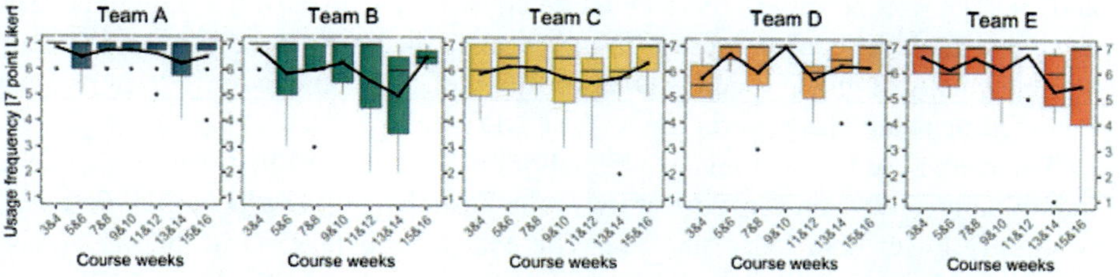

Fig. 5.4 Team level use of video conferencing tools

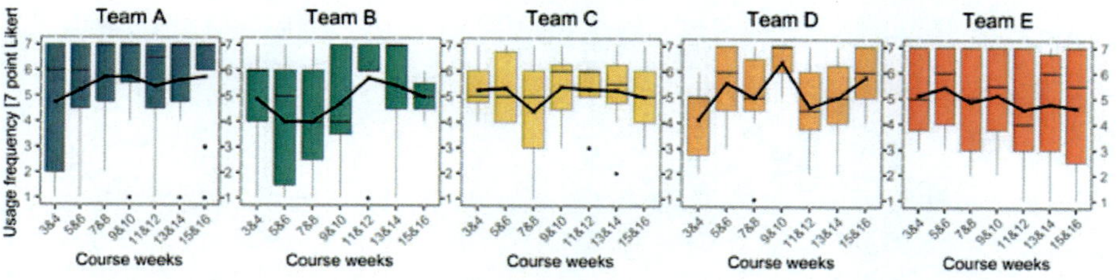

Fig. 5.5 Team level use of instant messaging tools

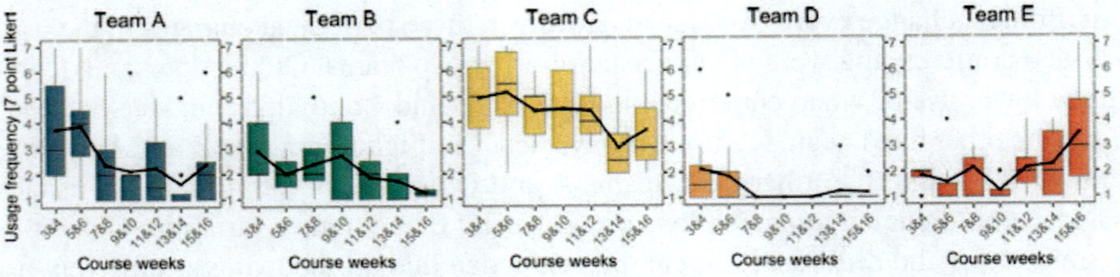

Fig. 5.6 Individual use of task management tools across teams

very low frequency of using the task management tools and finished almost without using them.

The collaborative use of task management tools (Fig. 5.7) usually aligns with the individual use (Fig. 5.6). However, teams A, B and C used task management tools slightly more in collaborative mode than in the individual mode, with the exceptions of teams D and E. As team D did not report using task management tools at the individual level, it might be that they also did not use them collaboratively. On the other hand, team E had different dynamics (increasing trend rather than decreasing) on both the individual and collaborative working modes compared to other teams. Furthermore, this team also used task management tools more often in individual mode than in collaborative mode, differing from the rest of the course teams.

The individual use of CAD tools was mainly in the later phases (Fig. 5.8). While team B started earlier (course week 7), teams A and C started later with using the CAD tools (course week 11). Teams D and E increased individual CAD use in between, at course week 9. Furthermore, most of the teams (except team D) decreased the use in the last two course weeks (i.e., weeks 15 and 16) after reaching a peak in weeks 13 and 14. The exception was team D which constantly increased the individual CAD use throughout the course, thus reaching a peak in the last two course weeks. Finally, while teams A, B and C used CAD uniformly with very little variation, teams D and E sometimes had a large variation in the CAD use.

The teams used CAD tools also collaboratively (Fig. 5.9) in a similar manner as individually. However, for most of the teams (except team D), the collaborative use is slightly lower than individual use. The exception is team D in the later phases (weeks 11–16), where they had a similar distribution of individual and collaborative use. Like the individual use, team D consistently increased the collaborative CAD

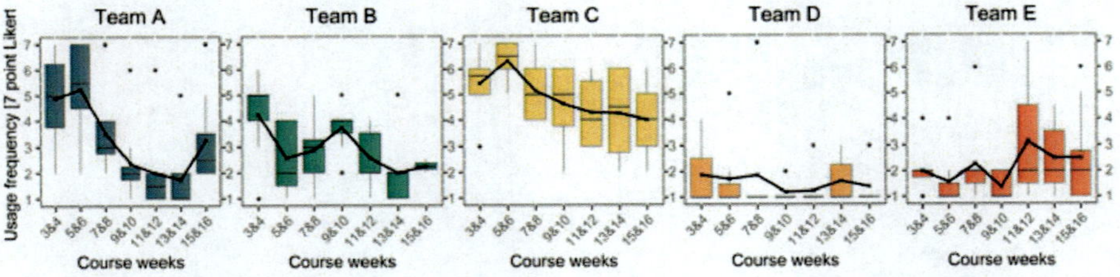

Fig. 5.7 Collaborative use of task management tools across teams

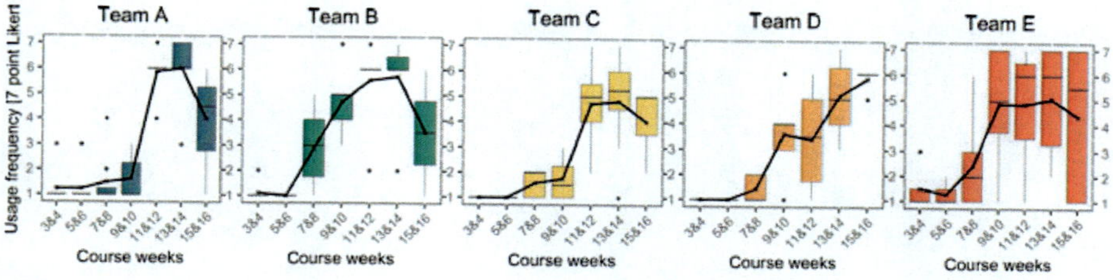

Fig. 5.8 Individual use of CAD tools across teams

work as the course advanced, having the peak at the end of the course. The same curve shape between individual and collaborative CAD work is in team C, which decreased collaborative CAD use only in the last two course weeks. However, teams A, B and E decreased the collaborative CAD use slightly earlier than individual work, which results in the earlier collaborative CAD work peak (weeks 11 and 12) for these teams. These results suggest that teams A, B and E focused more on the individual CAD work in the later modelling phases. Variation among team members is also visible in some course weeks. For example, all the teams had more variation in their peak time for collaborative CAD modelling (Fig. 5.9) than in the peak time of individual CAD work (Fig. 5.8).

Teams differed a lot in the use of file repository tools (Fig. 5.10). Teams A and D had very low use, team C used them a lot, while teams B and E were in between. While teams A, D and E were consistent using a file repository throughout the course, teams B and C differed substantially in their use. Team B started low with the file repository use and gradually decreased their use until weeks 9 and 10. After these weeks, team B again started to use a file repository more often. On the other hand, team C started high with the use, which gradually decreased until course weeks 9 and 10. After these weeks, team C had similar use to team B until the end of the course. Furthermore, teams A, C and D usually had lower variation in the use of the tools than teams B and E.

The individual use of visualisation tools usually has a form of the shifted decreasing curve (Fig. 5.11). However, the timing of the change from uniform usage to descreasing usage varies among teams. More specifically, team B has an early shift at the course week 5, team D has the late change at the course week 13, while the rest of the teams have the shift in between, i.e., teams A and E have a shift at week

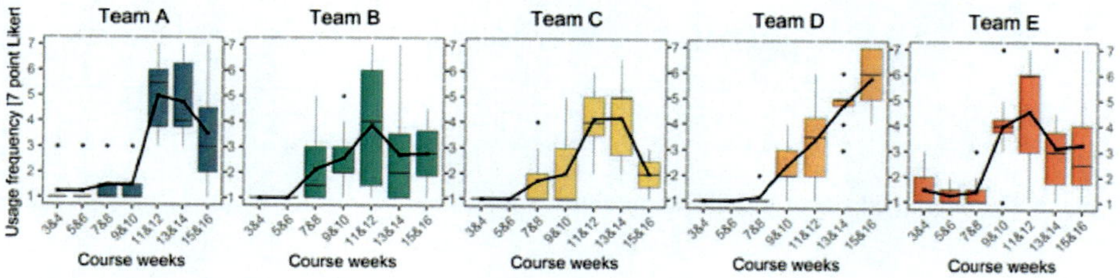

Fig. 5.9 Collaborative use of CAD tools across teams

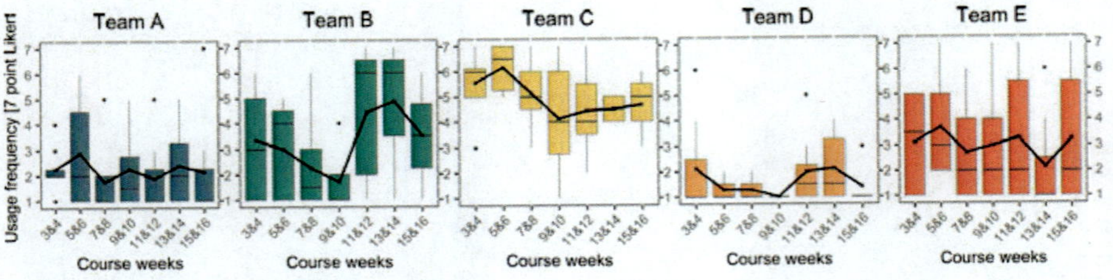

Fig. 5.10 Individual use of file repository tools across teams

11 and team C at week 9. For most of the teams (except team B), the usage peak was during the conceptual design phase (weeks 5–9). The use in the later phases varied among teams, as teams A, C and E used the visualisation tools in the last four weeks (13–16), while teams B and D barely used the visualisation tools individually. As with the other tools, differences between teams exists in the usage variation of their members. More specifically, teams B and D had consistently lower variation in the individual use of visualisation tools as compared to teams A, C and E.

Collaborative use of visualisation tools (Fig. 5.12) was similar to the individual usage mode. However, for most of the teams, collaborative usage was slightly higher than the individual one. For most teams (except team B), this difference is especially emphasised in the usage peaks. Contrary to the difference in individual and collaborative usage of CAD, where the peak points differed, all teams except Team C had aligned peaks between individual and collaborative use. Variation among team members is also visible in some course weeks. For example, all the teams had lower variation in their peak time for collaborative use of the visualisation tools (Fig. 5.12) compared to their variation in the peak time of the individual use (Fig. 5.11).

When taken together, these data suggest that teams do not differ much in their usage of ICT tools. However, some differences still exist, and they should be analysed more in detail. Furthermore, the results show that some tools support more early course phases, while others support later course phases. In addition, some tools are used throughout the course and support both early and later phases of the PBL design course.

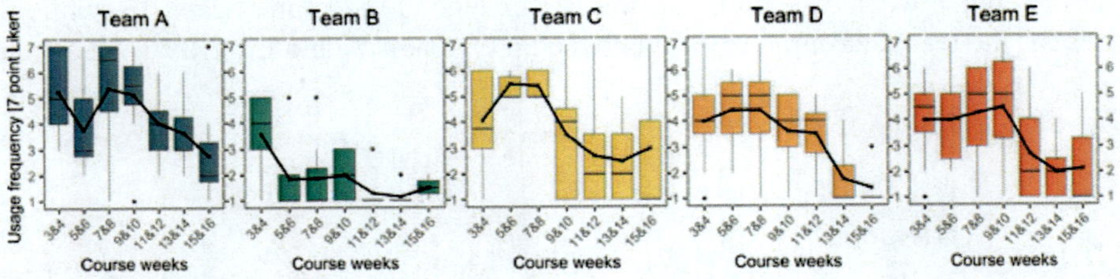

Fig. 5.11 Individual use of visualisation tools across teams

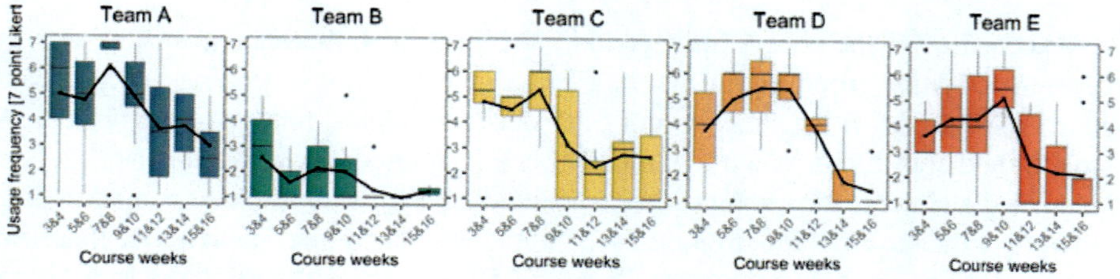

Fig. 5.12 Collaborative use of visualisation tools across teams

5.5 Discussion

All the different types of ICT tools explored in the longitudinal study show that the functionalities they offer become relevant in at least one phase of the project, if not more. Moreover, Figs. 5.2 and 5.3 display that the general trends of individual and collaborative use of ICT tools (at least for those ones that enable both the interactions), follow a similar pattern during the phases of the project and more specifically across the seven sampling points considered for the investigation.

For the discussion of the above results, it is important to start from the analysis of usage of the file repositories, as they are the only ICT tools considered here that can be exploited just from an individual perspective. The use claimed by the students is generally constant during all the phases and relatively low across the project, when the respondents are considered all together. On the other hand, there are larger differences if the analysis is carried with a team-wise perspective. The common trait among all of them is the scarce use of file repositories during the second phase of the project, when they were asked to generate the concept to further develop and embody in the next steps. This could be strongly related to the peak of use of the visualization tools for shared collaboration (e.g. Miro), which the students demonstrated to use with more significant intensity in the same time frame. File repositories, in fact, collect items that have a more strongly formalized degree of representation, as each file can contain contents according to a very specific structure: a CAD model contains 2D/3D geometry, a spreadsheet mostly collects numerical or categorical data; while text files might deal with figures and text, but still with the structure of a page. On the contrary, visualization tools that are already conceived for collaborative use enable the collection of multiple elements at the same time in a less structured way: they can hold text, tables, figures, numerical and categorical data, and all of them can be seen at one glance. The 2D/3D geometry of CAD models can be displayed as pictorial representations/images, although they are still not accessible for editing within the frame of these tools. Therefore, the opposite behaviour in terms of usage frequency of file repositories and visualization tools in phases two and three might be the consequence of how the communication requires the media to change during the project, as the design holds different types of information as long as the project proceeds.

According to the above findings and coherently with the general structure of the design activities proposed during the PBL course, it is not surprising that the teams used CAD tools with a stronger intensity during the latest phases of the project. As mentioned above, they can find it mostly beneficial once they have, at least, a general idea of the concept they want to develop and the way it should be embodied in order to deliver the required functionalities. Nonetheless, it is important to notice that some teams decided to start working with the CAD tools earlier than others (e.g., in the middle of the conceptual design stage). This can happen for two different reasons. First, some students within the team might find it difficult to explain their own ideas to the other teammates by means of more abstract representations that are typical of the initial phases and might decide to provide the others with a more concrete visualization of their concepts. Second, this could be due to the need of some teams to run some early verification of the feasibility of their concepts, as just a concrete embodiment of the same could do. These hypotheses appear to be the most likely to take place during the execution of the design PBL course, but it is yet unclear what is the real reason behind this behaviour, which should be studied with a future, dedicated investigation.

All the above considerations lead to the analysis of the elements (Becattini et al., 2020; Horvat et al., 2021) that made the delivery of information during the project efficient and effective. In fact, the tools that enable the collaboration cannot avoid the communication element, as the most used tools across the different phases of the project are those that enable videoconferencing and instant messaging. They show constantly medium–high values across all the phases and across all the sampling points, despite some teams had more evident fluctuations than others.

The video conferencing tools are the ones that the students used most frequently and intensively across all the phases of the project. This result could be also due to the more frequent use of these systems that all the partner universities involved in the research made for their standard lectures that had to be held remotely due to the conditions of the COVID-19 pandemic. At the same time, this result also confirms that the increased familiarity with a tool might have beneficial effects on its adoption (Horvat et al., 2021; Verstegen et al., 2016). One other possible explanation for this intensive use of video conferencing tools might be the need of meeting the others and mimic in-person live interactions to share information in the most effective way that was possible. From this viewpoint, it would be interesting to deepen the investigation in the future and explore which features the participants used most between the audio calls enriched with the screen sharing option and the video calls, which could be held both with and without the screen sharing option on.

The tools for instant messaging show a very similar trend, with a medium–high usage by all the participants and across all the teams considered for the investigation and, on average, a constant degree of use through all the phases of the project. Some teams showed peaks of use in different phases, where others had a less intense messaging activity. This might be justified by a different communication style that can characterize the five teams that participated in the research. Teams were composed homogeneously with people from four universities with a very similar background, but individual exigencies during the semester might affect the final outcome. In addition, the familiarity of the team members with the chosen instant messaging tool for the design PBL course could also have an impact on this result. In such cases, the exploration of the dynamics within the team should go through a more fine-grained characterization of the individuals, as this could provide additional factors that could explain these differences. The larger dispersion of data in specific phases of the project that takes place in specific teams could also reflect that some members have specific exigencies during the project. On the other hand, the dispersion of data at specific sampling points could also be due to the nature of the instant messaging tools (Dennis et al., 2008). In fact, compared to the video conferencing ones, these tools do not require a fully synchronous involvement of all the team members (or at least a large majority of them).

Tools for task management show an overall decreasing behaviour as long as the project activities proceed, without particular differences between the individual and collaborative use. However, the behaviour across teams is not as uniform as seen from a comprehensive perspective. Three teams out of five used project management tools at the start of the project with a marked intensity, highlighting a relatively easy adoption of the same. Then, progressively, they decreased their use of these tools from both the individual and the collaborative perspective. These results might highlight one or more phenomena that took place at the same time. On the one hand, some teams might feel the need to address the project activities via project/task management tools because they were told about their existence at the beginning of the project and the educators stressed the need for appropriate project management in order to ensure an efficient and effective execution of the same. On the other hand, these results could reflect the change in the structuredness of the different project phases. While it is expected that the activities can be more easily managed once the design has already a more or less stable embodiment, the activities could require stronger management when the concept still has to be consolidated in its functional features. Contrarily from the above three teams, one of the remaining teams showed the opposite behaviour, with a slow start and a progressive increase in the use of task management tools (team E). Additionally, one team declared an almost absent use of ICT tools for task management (team D). This could suggest that different teams might have different project management styles. For example, some teams might benefit from more structured project management at the beginning of the project because they want to keep the phases that are more creative clearly organized and with a definite set of human resources allocated to different tasks. Others might like to keep these phases as free and unconstrained as possible, so that all the team members can take part in these activities and bring their own contribution even if this requires the member to keep discipline during the meetings. Furthermore, as potentially witnessed by the team that did not almost use project management tools, it is possible that these different behaviours reflect a different strategy for project management. For instance, some teams could have used ICT tools meant for other purposes to compensate the apparent inefficiencies of existing project management tools, just because the alternative they selected enabled easier integration of project management functionalities into a single application (e.g., Miro integrates the display of Gantt diagrams).

5.5.1 *Implications*

The study of the use of ICT tools explained and discussed above can have significant implications for three different kind of profiles that can benefit from such findings:

- educators in engineering design or researchers in engineering design education;
- engineering design professionals that work remotely in collaboration with a team; and, finally,
- ICT tools developers, especially those focusing on the provision of tools for engineering designers.

For what concerns educators as well as researchers in engineering design education, it is important to recall that the work by Albers et al. (2009) pointed out that it would be hard to capture the motivation by the sole investigation of the use of ICT tools during the project execution. However, the longitudinal perspective offered within this research and that is replicable in other studies, show that students remained highly engaged during the whole project duration, as witnessed by the strong use of video conferencing and instant messaging tools (Brisco et al., 2016; Mamo et al., 2015). From the perspective of the interplay between individual and collaborative learning, as highlighted by Brisco et al., (2020a, 2020b), the results captured with the ICT tools confirm that the proposed platform for geographically distributed design collaboration is effective because the general trend of interaction with the tool is coherent in these two scenarios. Nevertheless, it would be interesting to check the extent to which the collaboration is fostered by the self-learning compared to the beneficial effects of the shared environment, as the students could also be capable of compensating lacks and learn from their peers as well.

In addition, educators that would like to repeat similar initiatives for geographically distributed project-based learning should pay particular attention to the setting they propose to students, as this can have an impact on the effectiveness of their activities and, therefore, on their learning process. The results here exposed show that a seamless execution of project activities will require to implement an e-learning platform (Becattini et al., 2020) that facilitates data and information sharing with multiple means of communication, as there is no ad hoc solution so far. At the same time, the way educators introduce the tools composing the e-platform can have an impact on the effectiveness and the efficiency of the students' workflow (Horvat et al., 2021), as witnessed by the lack of usage of project management tools during the last phases of the project.

The professionals with a background in engineering design that have to deal with design projects in geographically distributed settings or that have to work remotely because of different restrictions could use the above results in order to make their workflow more efficient and to plan activities with adequate tools. While it is expected that most of their activities already leverage CAD tools as well as video conferencing and instant messaging ones, it is possible that they should still explore the benefits of shared spaces for virtual collaboration that facilitate idea sharing. The effects recorded with this longitudinal study show that these systems are mostly used during the early stages of the project activities, which in this case correspond to the conceptual design stage. However, it is necessary to underline that these systems could be used also to create annotations in a collaborative way, which could be, for instance, conveniently used in combination with screenshots of the 3D models they develop during the latest stages of the product development process.

The effects of the investigation carried out and discussed above might also be extremely relevant for the developers of ICT tools for design, especially CAD/CAE tools. In fact, the perspective considered here span through many phases of engineering design projects, from product planning/task clarification up to the embodiment of the solution. The variation in the use of the different functionalities offered by the ICT tools here considered show that the existing 3D CAD modelling software are not sufficient to cover effectively the whole project. Current 3D CAD systems actually enable an efficient management of the work starting from the embodiment design stage (Gaha et al., 2021; Leipold, 2020), with a general lack of support for the early stages of the product development process, where the data/information set to manage is less structured. In the light of the changing scenario that requires a more intense remote collaboration, as forced by the evolution of the working conditions set by the COVID pandemics restrictions and for which it would be hard to make a step back, the future CAD systems should enable a stronger integration with visualization systems for collaborative work. These are vital for projects that have a significant innovation target, such as the ones dealing with new product development or technological shifts. It is more likely in these design steps that co-designers need to share their ideas with schemas, conceptual maps and similar diagrams that facilitate the representation of abstract items (e.g. functions, working principles, which are needed to be defined well before structures) (Gero & Kannengiesser, 2004; Idrissov et al., 2020; Pahl et al., 2007).

Still on the side of the development of CAD tools, the strong use of video conferencing as well as instant messaging tools during the whole project could steer the development of CAD systems that embed options for chatting or audio/video calling during the editing of CAD models. This feature would be progressively more and more needed in remote and/or geographically distributed settings, with a wider arena of potentially interested users. In fact, A/V calling or chatting could facilitate both the early and the latest phases of the product development processes, as shown across the different phases of this longitudinal data acquisition. During the early stages these functionalities will complement the ones concerning visualization described above, while in the ones mostly dealing with the development of 3D geometry they could facilitate co-development as well as the execution of design review meetings. The recent integration of some of these functionalities in more advanced design platforms (e.g. Dassault 3D Experience) suggests that this should be an evolutionary trend for CAD systems to be monitored.

These implications open up avenues for further improvement of the way design project-based courses are delivered in physical, remote or hybrid conditions in post-pandemic period. Various insights obtained during the delivery of design PBL courses indicated that students used ICT tools for both individual and collaborative learning, independently from the restrictions of the COVID-19 settings. Courses which are delivered in distributed settings are, by definition, requiring similar support tools to those forcedly carried out during the sudden COVID-19 transition. These experiences at the academic level allow educators to implement similar ICT tools and activities in their regular courses. At the same time, if these approaches become more regular at higher education level there are possibilities for life-long learners to implement them in professional contexts.

5.5.2 *Limitations*

While providing insights for researchers and practitioners, the presented work has some limitations that need to be accounted for. The most important limitation is the nature of the data collected, which is self-reported and aggregated over the two weeks period. However, to overcome the biases of the self-reporting estimation, the experience of researchers who coached two teams has been used to check the validity of the data. Furthermore, some tools enabled analysis of the time spent while using them (e.g., Onshape), so these objective measures were used to provide additional support to the validity of the self-reported results. Next limitation relates to the context of the distributed design PBL course, as it builds on the proposed e-learning infrastructure (Becattini et al., 2020) with the initial introduction of the proposed tools (Horvat et al., 2021). Finally, the study is limited in scope as it provides insights on five teams with constant size and during limited duration (four months). The dynamics of using ICT tools and interplay between individual and collaborative use might be different in other scenarios.

5.6 Conclusions

This study presents a longitudinal study of the ICT use during a distributed project-based design course. The results show that teams use various ICT tools during the course. However, while the use of the tools follows similar trend for individual and collaborative working mode, some tools are used consistently throughout the course, while the use of others was related to the course phase. Furthermore, differences in the extent of using different ICT categories exists. More specifically, the use of video conferencing and instant messaging tools was uniform throughout the course, however, video conferencing is consistently used the most, while instant messaging is used slightly less. Visualisation tools are used more in the early course phases, while CAD tools in the later phases. The use of file repository varied most among teams, suggesting different working styles in terms of work sharing. Finally, teams usually gradually decreased in their use of task management tools as the course proceeded.

The study was conducted during the COVID-19 pandemic, thus having implications for both distributed work in usual and restricted circumstances. Educators should use the obtained results in order to organise introductory lectures and tutorials of the ICT tools required for the PBL design course. While most ICT tools are necessary from the beginning of the PBL course, some could be introduced later when their usage becomes more relevant. Furthermore, educators could compare and translate these findings related to individual and collaborative student work to their teaching context and explore differences and similarities. These newly obtained insights could provide educators with insights on setting up the e-learning environment for the project-based courses to facilitate collaboration and knowledge exchange. On the practical side, engineering design teams (of both students and professionals) could use the presented results to support their transition to distributed work in a more efficient manner. For example, they could start exploring the benefits of shared spaces that support early design phases. Finally, ICT tool developers could use the presented results to further improve their offerings by introducing and integrating new functionalities and features (e.g., visualisation and video conferencing within CAD tools).

Future studies should explore the other ICTs used throughout the PBL course, and map their usage to various design course activities. This would allow us to understand the complementarity and redundancy of their functionalities in addition to underlying mechanisms for their selection and utilisation during the course. The following study should also address relations between the patterns of the ICT tool use and generated design outcomes. Furthermore, other distributed course contexts should also be investigated in order to provide a generalisation of the findings.

Acknowledgements This paper reports on work funded by the Croatian Science Foundation project IP-2018-01-7269: Team Adaptability for Innovation-Oriented Product Development—TAIDE (http://www.taide.org) and Erasmus + project 2018-1-HR01-KA203-047486: ELPID—E-learning Platform for Innovative Product Development (http://www.elpid.org/).

References

Albers, A., Sauter, C., Maier, T., Geier, M., & Ottnad, J. (2009). Academic Engineering Design Education in a Realistic Environment. In *Volume 8: 14th Design for Manufacturing and the Life Cycle Conference; 6th Symposium on International Design and Design Education; 21st International Conference on Design Theory and Methodology, Parts A and B*, pp. 565–573. ASMEDC. https://doi.org/10.1115/DETC2009-87425

Asgari, S., Trajkovic, J., Rahmani, M., Zhang, W., Lo, R. C., & Sciortino, A. (2021). An observational study of engineering online education during the COVID-19 pandemic. *PLoS ONE, 16*(4), e0250041. https://doi.org/10.1371/journal.pone.0250041

Becattini, N., Škec, S., Pavković, N., & Cascini, G. (2020). E-Learning Infrastructure Prototype for Geographically Distributed Project-Based Learning. In *Proceedings of the Design Society: DESIGN Conference*. https://doi.org/10.1017/dsd.2020.282

Brisco, R., Whitfield, R. I., & Grierson, H. (2020a). A novel systematic method to evaluate computer-supported collaborative design technologies. *Research in Engineering Design*. https://doi.org/10.1007/s00163-019-00323-7

Brisco, R., Whitfield, R. I., & Grierson, H. (2016). Recommendations for the use of social network sites and mobile devices in a collaborative engineering design project. In *Proceedings of the 18th*

International Conference on Engineering and Product Design Education: Design Education: Collaboration and Cross-Disciplinarity, E and PDE 2016.

Brisco, R., Whitfield, R. I., Grierson, H., & Bohemia, E. (2019). Overcoming the challenges of global collaboration through design education. In *DS 95: Proceedings of the 21st International Conference on Engineering and Product Design Education (E&PDE 2019)*. University of Strathclyde, Glasgow. 12th—13th September 2019. The Design Society. https://doi.org/10.35199/epde2019.73

Brisco, R., Whitfield, R. I., Grierson, H., Howell, B., Unson, C., Matheson, C., et al. (2020). Developing an online course in Computer-Supported Collaborative Design (CSCD). In *Proceedings of the 22nd International Conference on Engineering and Product Design Education*. The Design Society.

Dennis, A. R., Fuller, R. M., & Valacich, J. S. (2008). Media, tasks, and communication processes: A theory of media synchronicity. *MIS Quarterly: Management Information Systems*. https://doi.org/10.2307/25148857

Dym, C. L., Agogino, A. M., Eris, O., Frey, D. D., & Leifer, L. J. (2005). Engineering design thinking, teaching, and learning. *In Journal of Engineering Education*. https://doi.org/10.1002/j.2168-9830.2005.tb00832.x

Gaha, R., Nicolet, P.-M., Bricogne, M., & Eynard, B. (2021). Teaching experiments for engineering education based on cloud cad software. *Proceedings of the Design Society, 1*, 2951–2960. https://doi.org/10.1017/pds.2021.556

Gero, J. S., & Kannengiesser, U. (2004). The situated function-behaviour-structure framework. *Design Studies, 25*(4), 373–391. https://doi.org/10.1016/j.destud.2003.10.010

Gopsill, J., Snider, C., Lia, E., Edgar, S. J., & Hicks, B. (2017). Automatic design structure matrices: A comparison of two formula student projects. In *Proceedings of the 21st International Conference on Engineering Design (ICED 17)* (pp. 31–40). Vancouver: The Design Society.

Grimheden, M., & Hanson, M. (2005). What is design engineering and how should it be taught?–Proposing a didactical approach. In *Proceedings of the International Conference on Engineering Design*. Melbourne.

Horvat, N., Becattini, N., & Škec, S. (2021). Use of information and communication technology tools in distributed product design student teams. In *Proceedings of International Conference on Engineering Design ICED21*. Cambridge University Press.

Idrissov, A., Škec, S., & Maier, A. M. (2020). Visualising systems: Mapping system features and interactive information visualisations in design. *Proceedings of the Design Society: DESIGN Conference, 1*, 2295–2304. https://doi.org/10.1017/dsd.2020.138

Kolmos, A., & de Graaff, E. (2007). Process of changing to PBL. In Management Of Change: Implementation of Problem-Based and Project-Based Learning in Engineering (pp. 31–43). Rotterdam: Brill | Sense.

Leipold, K. (2020). CAD for college: switching to onshape for engineering design tools. In *2020 ASEE Virtual Annual Conference Content Access Proceedings. ASEE Conferences*. https://doi.org/10.18260/1-2--34250

Leung, W., Wang, Y., & Kim, S. W. (2019). Global product development: Project-based multidisciplinary joint course. In *Proceedings of the 21st International Conference on Engineering and Product Design Education: Towards a New Innovation Landscape, E and PDE 2019*. https://doi.org/10.35199/epde2019.55

Lou, Y., & Kim MacGregor, S. (2004). Enhancing project-based learning through online between-group collaboration. *Educational Research and Evaluation, 10*(4–6), 419–440. https://doi.org/10.1080/13803610512331383509

Mamo, J., Farrugia, P., Borg, J., Wodehouse, A., Grierson, H., & Kovacevic, A. (2015). Using engineering design tools in multidisciplinary distributed student teams. In *Proceedings of the 17th International Conference on Engineering and Product Design Education (E&PDE15)*. Loughborough.

Pahl, G., Beitz, W., Feldhusen, J., & Grote, K. H. (2007). Engineering Design. *A Systematic Approach* (Third Edit.). London: Springer-Verlag London Limited.

Reymen, I. M. M. J., Hammer, D. K., Kroes, P. A., van Aken, J. E., Dorst, C. H., Bax, M. F. T., & Basten, T. (2006). A domain-independent descriptive design model and its application to structured reflection on design processes. *Research in Engineering Design, 16*(4), 147–173. https://doi.org/10.1007/s00163-006-0011-9

Robinson, M. A. (2012). How design engineers spend their time: Job content and task satisfaction. *Design Studies, 33*(4), 391–425. https://doi.org/10.1016/j.destud.2012.03.002

Robson, C., & McCartan, K. (2016). *Real world research.* Wiley.

Verstegen, D. M. L., de Jong, N., van Berlo, J., Camp, A., Könings, K. D., van Merriënboer, J. J. G., & Donkers, J. (2016). *How e-Learning Can Support PBL Groups: A Literature Review.* https://doi.org/10.1007/978-3-319-08275-2_2

Vukašinović, N., & Pavković, N. (2017). Use of virtual mobility to facilitate modern project-based NPD education. *International Journal of Engineering Education.*

Vukašinović, N., Fain, N., & Duhovnik, J. (2011). The influence of a company's strategy on creativity and project results in an NPD—case study. In *Proceedings of the 18th International Conference on Engineering Design (ICED 11)* (pp. 366–372). Lyngby/Copenhagen: The Design Society.

Chapter 6
European Global Product Realisation: Creativity and Innovation in Educating Engineers and Product Designers of 21st Century

Ahmed Kovacevic, Jozef Duhovnik, Imre Horváth, Dorian Marjanović, and Péter Horák

Abstract COllaborative DEsign in Virtual Environment (CODEVE) is a teaching methodology developed within the European Global Product Realization (EGPR) course over a number of years. Today's products are global and our students engage in their early professional practice facing challenges of working within distributed organisations to develop global products. Following early research on methods and tools in educating students for such challenges, the Global Product realisation course was initiated at the dawn of 21st Century and was performed since then as a collaboration between European Universities. Each year, an Academic Virtual Enterprise of participating Universities and an Industrial partner is formed in which students are distributed in international teams formed from multiple partner Universities. Educational activities and the project tasks are primarily communicated through video-conferencing and other synchronous and asynchronous means of communication. The design process model applied in CODEVE originates from the model of Pahl and Beitz, but is extended and adapted to suit the fuzzy front end of design projects performed in academic virtual enterprises. The extensions are related to creating a vision and implementing design research methodologies at the start of the project, blending phases of embodiment and detail design as well as bringing students for the first time in the final workshop which is aimed to culminate with the working

A. Kovacevic (✉)
City, University of London, London, England, UK
e-mail: a.kovacevic@city.ac.uk

J. Duhovnik
University of Ljubljana, Ljubljana, Slovenia

I. Horváth
Delft University of Technology, Delft, Netherlands

D. Marjanović
University of Zagreb Faculty of Mechanical Engineering and Naval Architecture, Zagreb, Croatia

P. Horák
Budapest University of Technology and Economics, Budapest, Hungary

D. Marjanović et al. (eds.), *Design Research: The Sociotechnical Aspects of Quality, Creativity, and Innovation*, https://doi.org/10.1007/978-3-031-50488-4_6

prototype and public presentations of the products. The CODEVE methodology was tested on projects which include design of consumer products, service driven products and industrial machinery. The evaluation of the methodology was supported by the Erasmus + funded project called Networked Activities for Realization of Innovative Products (NARIP) from 2015–2017.

The CODEVE teaching methodology enables students to work on an industrial project, it encourages them to understand and explore methods from other disciplines and helps them to overcome barriers of distributed environment. Similarly, they realise that communication style, relationships with teammates, and the availability and clarity of shared information play a crucial role in the realisation of the project.

The CODEVE methodology has been implemented in academic institutions in Europe and tested in both European and transatlantic projects with Universities from Europe and America. This chapter outlines advantages and challenges in conducting this type of educational projects including the influence of the selection of product, industrial partners, marketing, implementation etc.

6.1 Introduction

In this rapidly changing world, the future of many companies depends on globalisation of design, manufacturing, servicing and sales. A study published in March 2006 (Spinks et al., 2006) outlines an industrial view on what engineers who will operate in this century should be. The main message of the report can be summarised as; *"… At the heart the defining and enabling skills that form the core competencies of the engineering graduate… Three roles are identified. Firstly the role of engineer as specialist … Secondly, the engineer as integrator reflects the need for graduates who can operate and manage across boundaries, be they technical or organisational, in a complex business environment. Thirdly,…the critical role engineering graduates must play is providing the creativity, innovation, and leadership needed to guide the industry to a successful future. This is a vision of the future that underlines the vital importance of undergraduate engineering education to the UK engineering industry…".*

Two distinctive views on the development of these competences can be identified. The first, often referred to as the reductionist view, assumes that design competence is nothing other than a set of basic design abilities typically addressed individually. The opposite is the holistic view, which sees design competence as a synergetic construct of generic human capacities, as explained by (Horváth, 2006). Various authors argue that design competences are built in different contexts (Bourgeois, 2002). In the past, the emphasis was put on getting basic knowledge for a designer to possess and use. At that time, students were taught in a way which helped them to pass examinations rather then to solve successfully real life design problems. Recently, however, design problem solving capabilities have been given growing attention and various aspects of design competence have been investigated and addressed. Many authors analysed which industrial and pedagogical requirements of competences students should have and how to obtain these in university engineering design courses. Munch and Jakobsen (2005) identified the three most important characteristics of competence namely, contextual, behavioural and problem oriented. They argue that there is no universal deliverable for engineering design education but rather that specific design know-how should be conveyed to students depending on the goals, content and form of a design.

The competence development is normally assessed in terms of its operation to enable design problem solving. For instance, (Crain et al., 1995), put these in categories such as teamwork, information gathering, problem definition, idea generation, evaluation and decision making, implementation, and communication. The authors claim that these should be developed within introductory design courses and suggest that other competences are to be addressed in higher design courses to suit specific disciplines. In all cases, knowledge remains important, but it is more often considered as an element of engineering design know-how, rather than as the only goal of design education. Overbeeke et al. (2004), identified nine competences that need to be developed by industrial design engineering education, and grouped these as core and meta competences. Horváth (2006), analysed the connection between personal know how and that contained in a community of professionals. Berge et al. (2002) concluded that communal competences are becoming more important for industry nowadays. Typically, communal competences are multi-disciplinary collaboration, dislocated communication, balanced comprehension, and resource sharing, while personal competences are creativity, communication, integrative thinking, problem solving and learning from examples.

The organisers of the European Global Product Realisation (EGPR) course did recognise the importance of the above requirements and hence adopted and followed a holistic view on engineering design education. A comprehensive review of the research performed during the previous courses on the development of holistic design competences is reported in (Horváth, 2006), Based on the experience and publications, the organisers of the course adopted the view that design competence is a combination of five capacities. These are knowledge, skills, capabilities, attitude, and experience, as shown in Fig. 6.1. They are all strongly connected to provide the intelligence, knowledge basis, and problem solving capabilities required for solving real design problems.

Design knowledge relates to all subjects required for problem solving. This may be either related to or independent on the problem at hand. Design skills are learned abilities to perform a design action or execute a process. Both of these result from experience. Design capabilities are required to perform a function; attitude is a way of thinking, while experience is acquired through actual observations of solving practical problems. All five capacities should be equally emphasised in the educational programs in order to develop design competence in future engineers and designers.

This paper will present the methodology of methods applied in the EGPR course and address important issues observed during the years of performing EGPR course and reflect on competences acquired by the students.

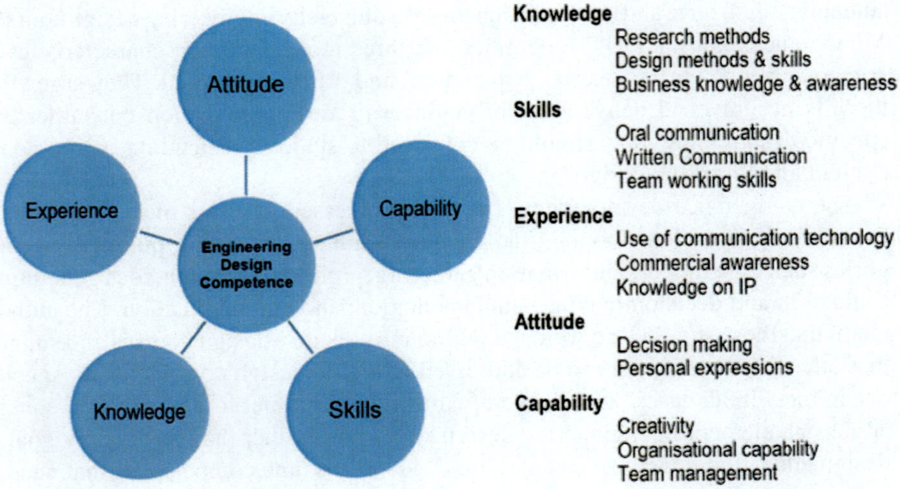

Fig. 6.1 Engineering Design Competence

6.2 History and Philosophy of E-GPR

European Global Product Realization course (EGPR) originally started as Global Product Realisation (GPR) by TU Delft, the Netherlands, University of Michigan, USA and Seoul National University, Korea in year 2000. It ran for two years but due to lack of tools for distributed synchronous communication and time differences between three continents was converted into a European project in 2002 (Horváth et al., 2004a, b), TU Delft, EPFL Lausanne, and University of Ljubljana joined to form the first project with the Slovenian company NIKO. Three more universities joined later, namely University of Zagreb in 2003, City, University of London in 2004, and University of Technology and Economics Budapest in 2009 (Kovacevic, 2016). In 2014, four European universities launched a joint educational project called NARIP (Networked Activities for Realization of Innovative Products). The project was supported by ERASMUS + funding (Vukasinovic, 2017). The history of University participants on the program is shown in Fig. 6.2. The project goal was to formalise, test and consolidate the methodology for collaborative new product development in a distributed environment by use of virtual tools.

In brief, each year participating Universities and an Industrial partner form an Academic Virtual Enterprise, as shown in the example from year 2018 in Fig. 6.3. Students are distributed in international teams formed from multiple partner Universities. The main communication comprising educational activities and project tasks is preformed through video-conferencing and other synchronous and asynchronous means of communication. The design process model applied in CODEVE originates from the model of Pahl and Beitz, but is extended and adapted to suit the fuzzy front end of design projects performed in academic virtual enterprises. The extensions are related to creating a vision and implementing design research methodologies at

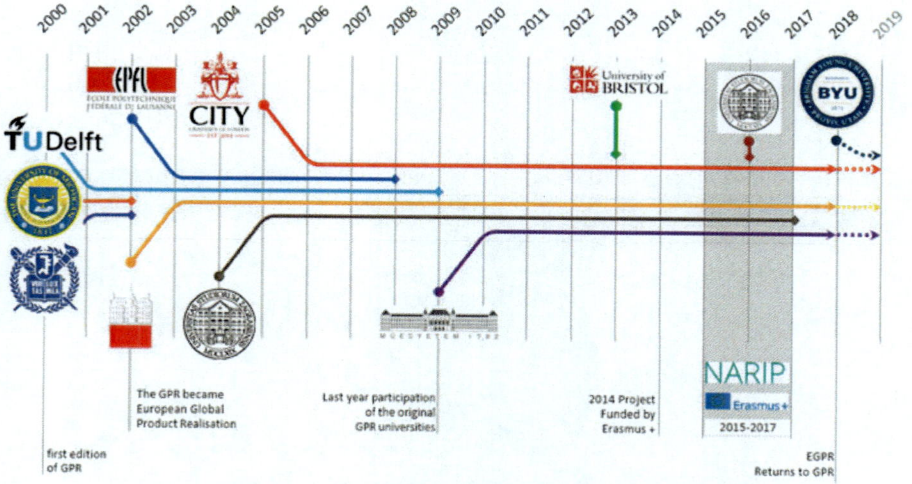

Fig. 6.2 Timeline showing milestones and University participants in the European global product realisation course

the start of the project, blending phases of embodiment and detail design as well as bringing students for the first time in the final workshop, which is aimed to culminate with the working prototype and public presentations of the products. The students are encouraged to perform conducted navigated active learning and include operational research in design process. At the end of the project, a hybrid prototype is assembled which often allows demonstration of IP generated for company.

The development of the teaching methodology was named CODEVE (Collaborative Design in Virtual Environment) and is explained in detail by Vidovics et al. (2016). The objective of the course is to expose students to effective methods in designing innovative products inside a distributed, collaborative, multidisciplinary, multinational and multicultural environment (Spinks et al., 2006) A wide variety of different projects with industrial partners have enabled a collection of broad and valuable insights and experiences over nearly two decades. The projects are unique each year and come from a variety of industrial sectors. They vary greatly in complexity, research and implementation as described by Pavkovic et al., 2011) and Kovacevic et al. (2016). The overview of the projects and partners participating in the EGPG course since 2008 is give in Fig. 6.4.

In 2017, the students' experiences in realising the NARIP project were summarised to evaluate suitability of the CODEVE teaching methodology for different disciplines and types of projects ranging from industrial design to engineering design. Tasks to design large industrial devices, like the welding inspection device for nuclear reactors from 2015, require a number of student groups to work on subsystems of a common prototype. On the other side, consumer products such as 2016's lighting solutions for aging population and 2017's lightweight mobility scooter require each student group to design and manufacture their own prototype. The first type of project is focused on engineering design while the second one leans

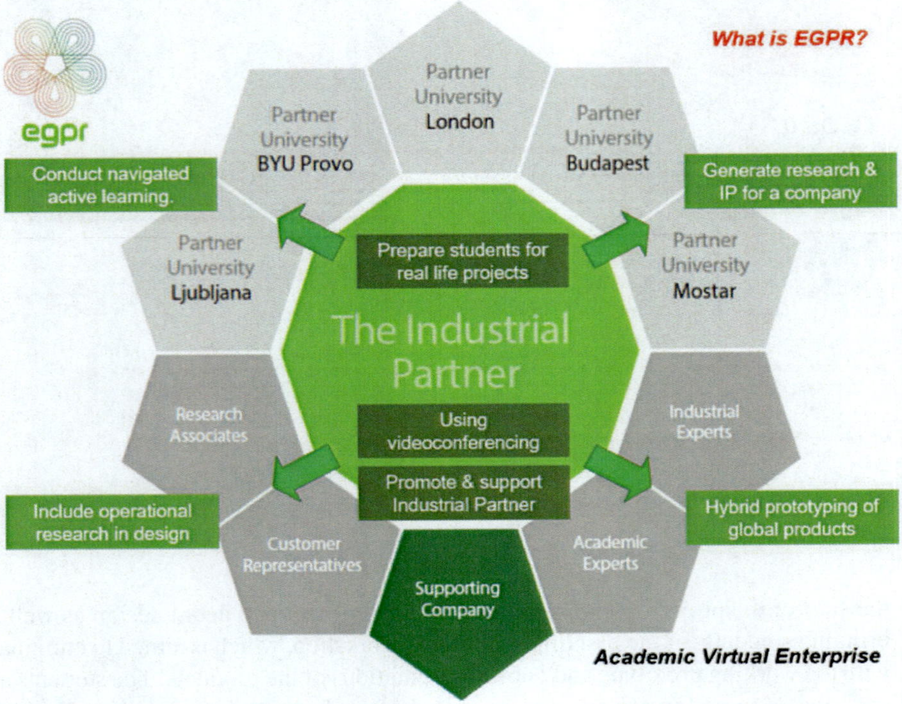

Fig. 6.3 Academic virtual enterprise of EGPR in 2018

towards industrial and product design. As shown in Kovacevic et al., (2017), it was confirmed that this teaching method was suitable for both and was ready for implementation in European collaborative projects. In 2018, a new partner, Brigham Young University from Provo in Utah, USA joined the EGPR community. Moreover, this year's industrial sponsor Black Diamond is based in Salt Lake City in Utah, USA and is a leader in outdoor climbing and skiing equipment. The project is hosted by City, University of London, marking the first time in the history of EGPR the partner company and the host university are not from the same country.

The next chapter will present a review of the effectiveness of the CODEVE methodology in this transatlantic project and expose the strengths and weaknesses of this methodology applied to teams consisting of industrial design, product design and engineering students collaborating within a globally distributed academic virtual environment.

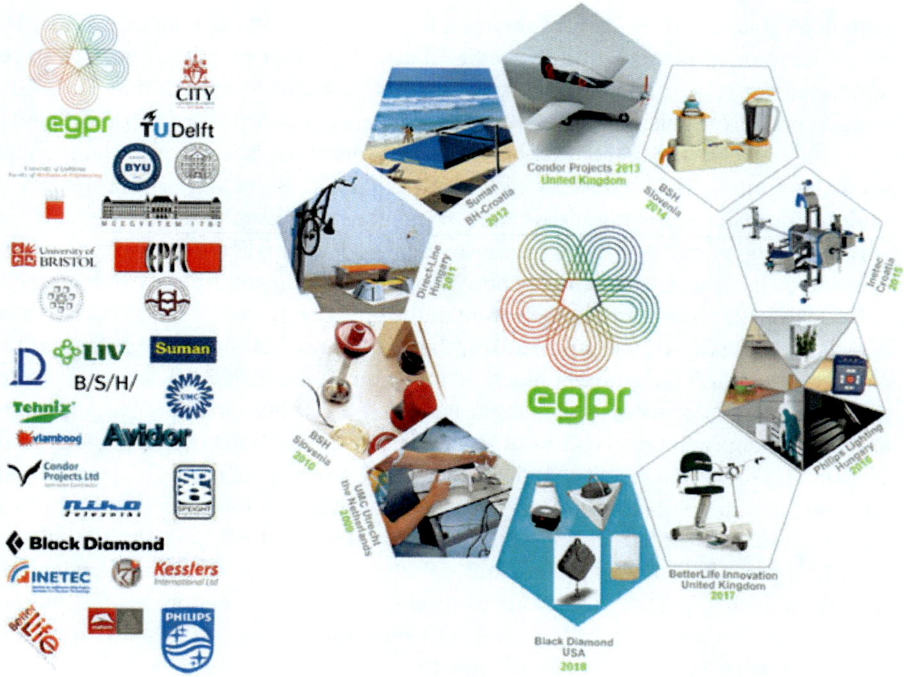

Fig. 6.4 Project realised in the EGPR course since 2009

6.3 CODEVE Methodology

Research in design and engineering education has shown that the traditional engineering design practice is not sufficient anymore, as it cannot face and satisfy all the new design requirements within a reasonable design time frame. Collaborative design is emerging as a promising alternative to classical design approaches. Teams of students with multi-disciplinary, multi-national and multi-cultural background are formed to enable an in-depth view of design problems. Various institutions are participating in the concept-to-market design process, making it even more complex. Furthermore, the nature of teams has changed significantly because of changes in organizations and the nature of the work they do. These new conditions of the business environment, being rooted in globalization, the explosion of new technologies, economy based on knowledge, and the information era have made working in virtual teams a common approach for many organizations today. Higher education is not necessarily aware of the respective emerging knowledge, skill, or competence requirements, and which may not currently be satisfied. In particular, the challenges of student projects being carried out in virtual teams in remote collaboration need to be addressed, because these projects are not parts of the traditional designer curricula. All these issues challenge the HEIs to be able to adapt to this paradigm change in design setting, and also to satisfy the emerging and changing knowledge, skill, and competence needs of the current situation (Crain et al., 1995; Kovacevic, 2008). The above mentioned theoretical issues as well as many other practical ones have been addressed in a series of international product development courses EGPR. The EGPR course came to existence as an answer to the concept of borderless education as well as to the major trends in digitally-supported design such as (i) design across value chains (globalization of product development, realization and marketing), (ii) design across multiple domains (growing importance of integrated multi-disciplinary design), and (iii) designing across life cycle processes (from conceptualization, through production and utilization, to recycling). These are indicating the multiplicity of the aspects to be dealt with, the multi-faceted nature of the knowledge the students need to learn, and the complexity of the problem from an educational point of view. The professional content and didactic approach of the course were designed accordingly; the course applied two instructional streams, which are called professional navigation and industrial project, and followed a generic four-phase NPD model (Spinks et al., 2006). The series of lectures and presentations are provided for all participating students, and the industrial project is carried out in 5 or 6 international, multidisciplinary virtual teams, all working on an industrial assignment given by the selected industrial partner. From the project kick-off all parties communicate and collaborate by virtual means, yet the product realization (prototype fabrication and testing) and presentation is done at the site of the host university in the frame of a week-long workshop, where participants finally meet in person.

As it has been described previously, the know-how and methodology in this project based design course for collaborative new product development (NPD) in dislocated, virtual environment went through significant development and participating institutions and individuals gained a lot of knowledge and experience throughout the years. Therefore, CODEVE is definitely not without antecedents.

CODEVE methodology is indeed a refined and crystallized know-how to set up and successfully manage a NPD student project in industry-academia setting in a dislocated environment. The CODEVE methodology is the primary output of the first project year in the NARIP Erasmus + Strategic partnership project.

The research and methodology development activity here was three-fold. Firstly, the recent and latest experiences both with NPD and virtual collaboration in the partners' practices (mainly related to the EGPR) had to be studied and processed. Secondly, the state-of-the-art methodological developments had to be discovered and the possibilities of effective implementation had to be identified. Upon the findings and conclusions, and also on the niches found, a streamlined approach and methodology applicable in virtual environment was formulated. Thirdly, the models were tested and continuously adapted to design education in virtual environment. For this purpose, an experimental industry-academia project was carried out (i.e. the NARIP EGPR student project), which was the subject of seeking and finding the most critical points for further development both in theory and in practice.

6.3.1 Design Process Model

The design process model applied in the project originates from the model of Pahl et al., (1995) but in an extended, adapted version. The first phase, depending on the type of project may depart from Clarification of the task, and become more of a Fuzzy Front-End (FFE)-type of problem definition. Once the product is defined in terms of the demanded functions (and further requirements), teams could enter the concept generation phase. Another difference from the Pahl-Beitz model may be that there is no separate design phase for embodiment and detail design, with no intermediate review. The third major adjustment is that there is a prototype making phase at the end. Eventually, the design process resembles more closely the whole product development phase in the innovation model of Roozenburg and Eekels (1995). In the course methodology there are a number of guidelines and written aids available to ensure a common understanding in terms of the design process to follow. The goals, recommended tasks, and also expected outcomes and deliverables of each phase are prescribed in details. This, however, does not mean that the designers would be limited by obligatory methods and tools; in contrary, only the meeting points are defined to ensure the comparable outputs in time and depth, otherwise students are free to decide which way they choose. The process of developing and realising ideas in CODEVE methodology is shown in Fig. 6.5.

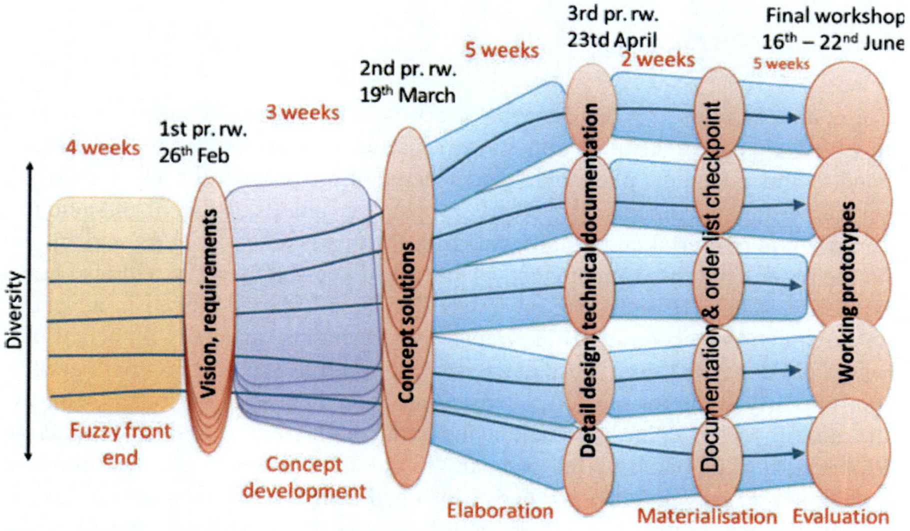

Fig. 6.5 CODEVE design process

6.3.2 *Teams*

In order to best simulate a real-life situation, the virtual enterprise of NARIP/EGPR acts like a flat-hierarchy virtual company, where the R&D and design departments are the student teams. The partner company is the customer, whereas academic staff takes only some higher level management roles, and otherwise facilitate interaction between company and student teams. The members of the teams are set before the project starts. Other than that, the team is an autonomous entity; it is responsible for setting up internal communication and working protocols, project and data management solutions, and definitely for the timely solution of the design assignment. Being a member of a dislocated international team, students might face challenges in language use or IT use, but most importantly, the greatest challenge is to actually perform as a team rather than eventually having the sum of individual efforts from remote locations. Within the team, not only the task distribution is important, but clear roles have to be set. This comes into focus, when the project assignment demands for a complex technical solution, where teams have to perform cross-team collaboration on top of internal team collaboration.

6.4 Project Preparation

6.4.1 *Partners*

For a successful project there has to be a sufficient number of partner universities involved, plus one industrial partner. As the partner company changes each year, they need to understand the philosophy and scheme of the project, for which there exist several written documents. In the early preparatory phases, the form and amount of contribution (material and immaterial) from company side has to be settled, while on the other hand the company expectations and possible benefits will also have to be clearly stated. IP rights are an issues that need to be addressed in advance as an agreement between industrial partner and organizing university on behalf of the whole project consortium. Further external, supporting and guesting partners could join the virtual enterprise, in the consensual agreement with the others. However, the most important contribution of the industrial partner is a document called Project Proposal. It is prepared by the company in collaboration with the host university, and in consensus with other partners. This document gives an overview of the aims and background of the project, briefly introduces partners, and most importantly the design challenge. The document specifies the project goals and expectations, recommends tasks to be performed by student teams, lists the deliverables with respective specifications, and also specifies phases, defining milestones with deadlines.

6.4.2 IT Communication and Collaborative Environment

The main means of communication and collaboration in a distributed environment are the Computer Supported Co-operative Work (CSCW) and Groupware solutions. Not surprisingly, the most widely used social platforms are utilised quite often to manage the teamwork. In terms of asynchronous collaboration in the course a few e-mailing lists are used, there is a shared workspace available for data exchange and backup. A whiteboard application is also available. The activities of joint problem solving (e.g. group ideation, common sketching, explaining and discussing the concepts, the discussion of needs for modification, common CAD modelling, etc.) are all still considered challenging, as even though the tools are available students may not be familiar with them. There has been a thorough document developed titled the "IT Guide", describing the official and optional IT solutions in details, furthermore there are chapters dedicated to proprieties and good practice in virtual environment.

6.5 Project Support

6.5.1 Academic Lectures and Professional Presentations

Although the EGPR project is aimed to be the final project for students before their employment in companies i.e. building on the already acquired knowledge, additional domain-specific lectures and topic-specific presentations are required to facilitate knowledge development. Academic lectures are delivered by renowned university staff, while professional presentations are held by external experts, professionals, and importantly, the representatives of the partner company. In terms of topics there are a variety of areas covered, e.g. project methodology and background, design methodology, relevant fields of engineering, management of virtual teams, CSCW solutions, creativity and innovation, presentation techniques, etc. In advance of the course start, the series of lectures and presentations are carefully planned in line with the logic and need of the current project and all are indicated in the Course schedule.

6.5.2 Coaching and Project Management

No project management can be successful without strict time management. The NARIP/EGPR timeline is specified with all details in a document called Course Calendar prior to the start of any student project activities. In this document, the course/project activities are broken down on a weekly basis. Two classes are scheduled every week via videoconferencing with defined titles, types of session, responsible location and a session moderator. Academic and professional lectures, student design review presentations, preparatory and consolidation meetings may use up the available timeslots.

The strict management of project and time is also crucial to synchronize the performance of the otherwise independent teams. Therefore, each virtual student team has a coach assigned (sometimes a co-coach as well), who is ideally an academic staff member with long coaching experience in student projects. The coach is essentially a point of reference; in the first place they enhance a common understanding in terms of tasks, duties, inputs, processes and the contents and form of delivery. The other major role of the coach is to monitor team activity and to point out underperforming or risks of failure well in advance. On the other hand, coaches and company representative in consensus with board of professors operationally manage the project. Coaches and company representative have regular weekly meetings (if necessary more frequently), to check the progress, evaluate the status against the work plan, and to analyze the possible risks on the level of the whole project. If necessary, these meetings can allow decisions to be made to initiate additional review points, prepare additional guidelines or protocols or apply shortcuts. This kind of continuous monitoring, quality control, and flexibility aims to realize the maximum effectiveness of all contributors and ensures that project goals are met successfully. In the project repository there are a number of documents and templates that can be used in different situations, however the management and quality assurance protocols are continuously evaluated and updated.

6.6 Project Closing

6.6.1 Closing Workshop

The project is 16–20 weeks long, and is divided into four phases according to the development process applied, each lasting 4–5 weeks, as shown in Fig. 6.5. The last phase, the Prototyping phase begins while teams are still operating in the distributed environment. It culminates with within the last project week (called the "Workshop week"), when all participants come together in the host country. The purpose of this co-located week is to assemble and test prototypes, and to present the project results to the academic staff, the company (generally located in the host country), and to a wider audience in a form of a large scale public presentation and exhibition. This is when the participants meet for the first time in person, which is always very motivating and a great experience. The peak point in the project is definitely the closing presentation and exhibition. This is a large scale event held at the host university campus. In practice, the closing day comprises of a series of presentation events. As EGPR is a university course, a formal academic-type presentation is required for final assessment and marking. A slightly different presentation is expected from student teams for company management with the emphasis adjusted to the interests of the audience. Testing of prototypes is carried out as a part of one of these presentations and it counts in the final assessment of students.

6.6.2 Scholarly Work

Throughout the years EGPR has provided a great opportunity to carry out experiments and research activities on each separate project. Besides having a distinct research focus in each year, the internal processes and phenomena were kept monitored by scientific quality methods. The latest findings and lessons learned are presented at relevant scientific conferences and journals. This activity serves dual goals; on one hand it significantly contributes to the quality assurance of the project, on the other hand it enables academics to extend their research work and research supervising activities both locally and within the EGPR community. After NARIP started, the approach has slightly changed. The main goal of the NARIP project was to consolidate and test a design education methodology for collaborative new product development in dislocated, virtual environment on variety of projects.

6.7 Projects Through Examples

The aims and objectives of companies collaborating on EGPR projects are different each year. This depends on the type of the business of the company, the sector in which the company operates, maturity of the company in terms of its position in the market and largely on the culture for NPD in the company. One of the objectives of NARIP and EGPR was to evaluate which type of product is the most suitable for this type of projects. Therefore.

6.7.1 Design of a Submersible Device for Inspection of Welds - Industrial Products

The NARIP 2015 was hosted by the University of Zagreb and the industrial partner INETEC - Institute for Nuclear Technology, both based in Zagreb, Croatia. For more than twenty years, INETEC has been a name synonymous for technological and service excellence in nuclear industry. They are active in research, development, design, construction and fabrication of equipment, tools, plugs and probes, including software and instruments for non-destructive examination.

In this project, students were faced with the challenging task of designing a remotely operated device for inspecting reactor pressure vessels in nuclear power

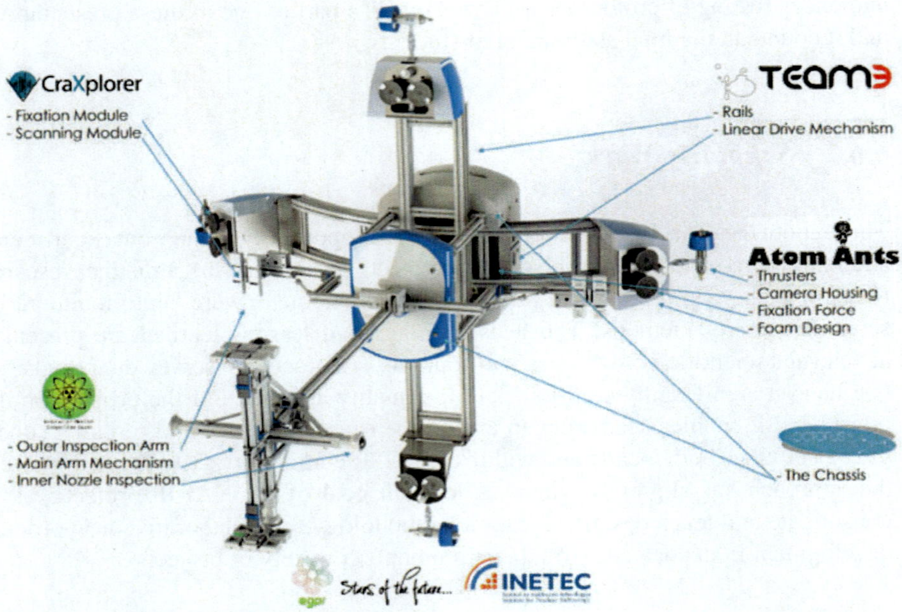

Fig. 6.6 Remotely operated device for inspecting welds in nuclear reactors, Zagreb, 2015

plants. Many aspects of the problem were investigated including; underwater propulsion, accurate location of vessel features, non-destructive testing methods and scanning procedures, power and data connections, and vehicle control. Total of 35 students from 4 universities were grouped in 5 international teams each focussed on a different subsystem as shown in Fig. 6.6. Students required excellent teamwork and communication in order to ensure compatibility between subsystems in the final prototype.

The week long workshop was hosted by the University of Zagreb in early July 2015. The assembly and testing of the single prototype was performed at INETEC facilities. The project demonstrated applicability of the CODEVE methodology for design of large devices for use in industry.

At the beginning of the project, students struggled to work collaboratively on this large device. This created a need for a cross-team and the update of instructions for different steps in CODEVE especially about the collaboration methods. Students used Conceptboard, an online whiteboard tool which proved to help both, in project management and communication.

Fig. 6.7 Design of intelligent lighting products for the ageing society, Budapest 2016

6.7.2 Design of Consumer Lighting Products in 2016 – Consumer Products

The host of NARIP 2016 was University of Technology and Economics from Budapest, Hungary. The project partner was Philips Lighting Hungary, subsidiary of Royal Philips of The Netherlands. The company is focused on improving people's lives through meaningful innovation in healthcare, consumer lifestyle and lighting. The project objective was 'Design of intelligent products for the challenges of the ageing society'. In this design assignment, the two most challenging areas are information sensing and processing related to visual and cognitive abilities respectively 39 students were grouped in 5 international teams. Each team developed their own vision which resulted in 5 working prototypes as shown in Fig. 6.7 (Vidovics et al., 2016). Prototypes ranged from the intelligent indoor gardening system to mood control lighting, heat detection system, intelligent stair lighting and stair climbing support systems.

6.7.3 Services Driven Products

The next type of products designed in this project are service driven products. The example is the project hosted by Technical University of Delft in collaboration with University Medical Centre from Utrecht in the Netherlands in 2009. Students had task to design devices for postoperative treatment of orthopaedic patients. Which will help patients and physicians in rehabilitation therapy. 32 students participated in the project and produced variety of solutions ranging from correcting posture of children to Wee technology to assess exercising and inform the consultant. As shown in Fig. 6.8, one of four groups designed product called Phoenix. It monitors the patient exercising at home. The docking station gives instructions on how to exercise. The patient puts markers on the specific locations on the body. The camera records the movement of the markers and the data is transmitted to the consultant

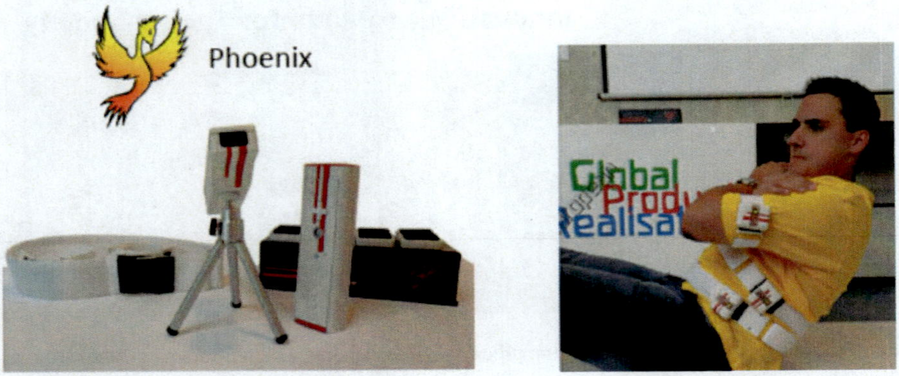

Fig. 6.8 Service based product UMC Utrecht, The Netherlands, 2009

over the internet. The consultant then can correct the patient and prescribe different exercises if required.

6.7.4 Fuzzy Front End Project

Most of the companies are open to innovation and allow students to start with completely open-ended projects. In the beginning, based on the project brief, students will create vision for personas of their choice and assess the social aspects of the product. Only in the later stages, with careful navigation from the company, the student will get in realisation of products and produce prototypes.

Figure 6.9 shows products and students participating in the transatlantic project in 2018 in which the Host University was City, University of London in the UK collaborating with Brigham Young University in Utah and BME from Budapest. The industrial partner was Utah based company Black diamond who asked students to design lighting solutions for outdoor activities. The students produced variety of interesting products.

6.8 Discussion

Surveys conducted in 2015 and 2016 established a benchmark for the analysis of projects in 2017 and 2018. The surveys were reasonably comprehensive and the full results were published in (Vukasinovic, 2017). Here we only present elements of the survey related to the project execution. In 2015 most students participated in the survey (33 of 35), while in 2016 only 30% of students returned the survey (12 of 39). The response to questions was given on the scale 0 to 5, 0 meaning 'no influence' with 5 meaning 'heavy influence'. Despite the relatively low response rate in 2016,

Fig. 6.9 Fuzzy front end project for black diamond, Utah, USA, 2018

the standard deviation for year 2016 was similar to that from 2015 and ranges from 0.5–1.1. Therefore the results were accepted as relevant.

Surveyed students felt that the level of achievement in 2016 when the project was about design of a relatively small consumer product is higher than for the industrial project performed in 2015. Students felt that the project objectives, target costs, and reduced complexity have been better achieved in 2016 when students worked one consumer products. This is probably because working on individual prototypes, students were able to have more control of the process and are more personally related to the final product. However, students felt that in both years the company goals were met but the achievement of company needs was overall lower than other individual criteria considered in this group of questions related to projects.

With regards to fulfilling the product specification set by students during the vision phase, it was shown that students who designed products for ageing population in 2016 were more satisfied with how their products matched specification. The complex product for industrial use designed in 2015 achieved only 70% of the target goals set in the vision phase. The consumer products in 2016 reached 90%. This shows that engagement of students in the project and their satisfaction with the results is better if such distributed design projects are related to consumer products. The supervision of the students and implementation of CODEVE methodology is also easier in this case. Although not subject of this study, the survey shows that it is easier for the academic staff to more directly implement CODEVE methodology for projects that address challenges of specific consumer groups through product design. The projects that focus on company engineering challenges are more challenging to realise.

The final part of the survey is related to how different factors affect student work in this international collaboration as shown in Fig. 6.10. Value 0 means no effect while value 5 means large effect. Results from both years are very similar. They show that the lowest impact on the project success is due to differences in cultural background of students. However the highest influence to students work and results is in the selection of design process and tools which are different in product and engineering deigns. For tasks related to consumer products, the selection of available tools is greater and the product design students can contribute more on aesthetics, ergonomics and user perception. Also, this requires engineering students to accept methods which they may not be using regularly in other design courses. Similarly, communication style, availability and clarity of shared information play crucial role in the realisation of the project. The improvements of CODEVE are possible in this area.

Another important factor is the availability of computer tools for implementation of CODEVE methodology. Nowadays, the tools for virtual communication are readily available and regularly used for social interactions and business. However not all of these tools are suitable for design projects and it is important to evaluate and improve CODEVE with respect to the new emerging communication technologies.

The next survey was performed in 2018. There were two reasons; firstly some changes were introduced in CODEVE especially in the fuzzy front end when the social aspect of the product design were emphasised and secondly because the project was performed with participation from USA and European Universities. In this study 53% of students (20 of 38) completed an online survey, in combination with randomized individual interviews with students. Results indicated that the process vocabulary differences between the different disciplines were more pronounced in 2018 than in

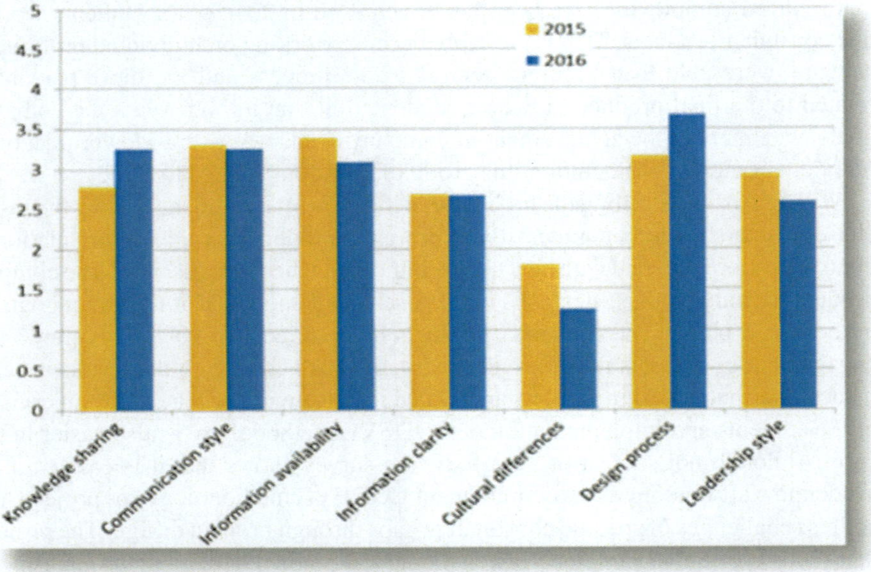

Fig. 6.10 Survey results for factors affecting team work using CODEVE

previous years when projects were organised within longer-term collaborating European University partners. While underlying goals were similar, there were frustrations as students tried to understand the vocabulary of other disciplines. Additionally, the general clarity of the fuzzy front-end methods and outcomes was low due to different starting times at three locations, London, Budapest and Utah with more than 3 weeks start between each university, causing issues transferring knowledge within teams. Documenting and presenting the work in different phases was also a challenge, as students are comfortable using virtual tools such as Google docs for asynchronous communication but are reluctant to use the blackboard-type system provided by the universities that allows monitoring of team progress. Varying methods of credit allotment between universities also caused stress as students discovered some disciplines valued certain phase components higher than the others, a phenomenon caused by deviance from the requirements of CODEVE methodology by the new-coming University. Finding common meeting times in 3 different time zones that are 7–8 h apart was also an enormous challenge; only one meeting with all participants took place in each phase. However, most students reported they either participated in or watched the majority of lectures and meetings as they were recorded and saved in the cloud for future viewing. Because students were distributed unevenly between universities, it is difficult to distribute tasks and follow the procedures evenly. Often team members from one university would meet and make decisions among themselves and neglect to share those decisions with team members in a timely manner, who continued operating on an outdated path.

A number of positive outcomes were also noted. The students enjoyed learning the processes and values of other disciplines and felt interdisciplinary collaboration creates more meaningful and complete products than individuals or single disciplines can. They also gained a respect for the challenges of working in different time zones, the importance of thoughtfully planning consequential communication, and the need to compromise and have patience with co-developers.

6.9 Conclusions

The framework of EGPR projects performed over 18 years by the Universities and companies from Europe and USA enabled development and demonstrated the applicability of the CODEVE methodology in a project based learning environment based on industry-academia collaboration. The teaching methodology for distributed Product Development courses presented in this chapter, illustrates details of the journey student take to achieve the desired final result of a new product development project—a full scale prototype, ready for testing and demonstration. CODEVE is not solely a university course description, nor simply an NPD methodology. From a different perspective it should be emphasized that this design course is one of a kind; here the R&D activities, the design and innovation processes and outputs are in good balance and just as important as the project itself, with all the project management considerations, the visibility of the project through the presentations and other PR activities, and the scientific publications.

Surveys conducted throughout the course showed the CODEVE methodology to be used for product design of industrial machinery and consumer products. Consumer related design projects are easier to manage and are more likely to meet the company project and product goals set by students through the 'voice of customers'. The selection of the design process, the communication style and the availability and quality of information are the most influential factors for the success of distributed design projects. New virtual tools are required for better implementation of CODEVE.

The re-introduction of transatlantic CODEVE project with participation of EU and USA partners was emotionally and cognitively polarizing, with students experiencing both elation and frustration with the course. The industrial and product design students were pushed beyond their traditional boundaries by including engineering practices that bring a product into a functional, operational reality. This will prove beneficial and distinguishing in their future employment applications. The engineering students were exposed to the values of a human-centred design process, the role of brand, and the importance of emotionally and functionally meaningful product designs, which will be equally useful for their future employment applications.

The CODEVE teaching methodology encourages students to understand and explore methods which they may not use regularly in their existing design courses. Similarly, communication style, relationship with teammates, and the availability and clarity of shared information play a crucial role in the realisation of the project. Such factors multiply the impact on student projects with participation from universities is different time zones, necessitating careful planning of process language and expectations, alignment of timing, simplification of tools and common understanding or phase deliverables for less dramatic transatlantic projects.

Acknowledgements The NARIP project 2015-2017 is a part of Erasmus+ Strategic Partnership program funded by the European Commission. This communication reflects the views of the authors, and the Commission cannot be held responsible for any use which may be made of the information contained therein. Authors would like to thank the European Commission for the support in realising and evaluating CODEVE Methodology.

The Authors would also like to thank all participants in this project since it's inception in 2002. This includes more than 30 companies, More than 30 staff members from 12 participating Universities and over 800 students who participated in the project.

References

Berge, Z., de Verneil, M., Berge, N., Davis, L., & Smith, D. (2002). The increasing scope of training and development competency. *Benchmarking: An International Journal, 9*(2), pp. 43–58.

Bourgeois, E. (2002). *Developing foresight for the development of higher education/research relation in the perspective of the European research area (ERA)*. European Commission.

Bufardi, A., Xirouchakis, P., Duhovnik, J., & Horváth, I. (2005). Collaborative design aspects in the European Global Product Realization project. *Interntionl Journal of Engineering Education, 21*(5), 950–963.

Crain, R. W., Davis, D. C., Calkins, D. E. & Gentili, K. (1995). "Establishing engineering design competencies for freshman/sophomore students. In *Proceedings of 1995 Frontiers in Education Conference*, ASEE-IEEE, pp. 4d2.1–4.

Horváth, I., Wiersma, M., Duhovnik, J., & Stroud, I. (2004a). Navigated active learning in an international academic virtual enterprise. *European Journal of Engineering Education, 29*(4), 505–519.

Horváth, I., Duhovnik, J., Xirouchakis, P., & Wiersma, M. (2004). Reflections of teaching global product realization in academic virtual enterprise. In *Proc.DETC'04 /57626, ASME*, Salt Lake City, 2004, p 3.

Horváth, I. (2006). Design competence development in an academic virtual enterprise. In *Proceedings of IDETC/CIE 2006 ASME 2006*, Philadelphia, Pennsylvania, USA.

Kovacevic, A. (2008). Competence development in an international product design course. In: Marjanović, D. Štorga, M. Pavković, N. & Bojčetić, N. (editors). Proceedings of the DESIGN 2012, 12th International Design Conference, University of Zagreb, Zagreb.

Kovacevic, A. (2016). *European Global Product Realisation.* Available: www.city.ac.uk/egpr. Accessed on 2018, 19 February.

Kovacevic, A., Vukasinovic, N., Pavkovic, N, & Horak,P. (2017). Evaluation of "CODEVE" Methodology for Teaching NPD to Virtual Design Teams. In: *Proceedings of E&PDE 2017*, Oslo and Akershus University College of Applied Sciences, Norway.

Munch, B., & Jakobsen, A. (2005). The concept of competence in engineering practice. In *Proceedings of Interntionl Engineering and Product Design Education Conference*, vol. 15–16 September, 2005, Edinburgh, pp. 1–8.

Overbeeke, K., Appleby, R., Janssen Reinen, I., & Vinke, D. (2004). Nine competencies, six units: Industrial design education at TU/e". In *Proceedings of International Engineering and Product Design Education Conference*, vol. 2–3 September, 2004, Delft, pp. 1–8.

Pahl, G., Beitz, W., & Feldhusen, J., & Grote, K.H. (2007). Engineering Design: A Systematic Approach. In: 3rd Edn, Translators and Editors: K. Wallace and L. T. M. Blessing, Springer-Verlag London Limited, 2007, ISBN-10: 1846283183.

Pavkovic, N., Marjanovic, D., Kovacevic, A., Fain, N. (2011). Industrial partnership in design education—experiences from EGPR course. In *Proceedings of E&PDE 2011*, pp. 8–9. London

Roozenburg N. F. M., & Eekels J. (1995). *Product Design:Fundamentals and Methods.* Wiley.

Spinks, N., Silburn, N., & Birchall, D. (2006). *Educating Engineers for the 21st Century: The Industry View.* A study carried out by Henley Management College for The Royal Academy of Engineering, avail. at: https://www.researchgate.net/publication/241751397_Educating_Engineers_for_the_21st_Century_The_Industry_View

Vidovics, B., Vukasinovic N., Pavkovic N., & Kovacevic A (2016). Development of methodology for distributed collaborative design environment. In *Proceedings of E&PDE 2016*, Aalborg University, Denmark.

Vukasinovic, N. (2017). NARIP, Available: http://narip.lecad.si/. Accessed on 2018, 19 February.

Chapter 7
Creativity—a Bottleneck in Engineering Design?

Check for updates

Udo Lindemann

Abstract Creativity is one of the important research topics in engineering design, industrial design, psychology and other domains. Researchers are looking for possibilities to improve the creativity of individuals or teams based on methods, tools, specific environment or mindset. On the other hand, it is well known that in industry just a very small percentage of ideas available are leading to successful products. Why should we generate even more ideas, as long as the success rate is that poor? Design Thinking claimed to address this topic; SCRUM and similar concepts of agile development did this in a different way. Are these just simple trial and error methods? Is the classical design methodology becoming obsolete? Combining all the valuable methods and concepts including LEAN lead us to the TMS- approach developed at TUM. Interdisciplinarity, lean, agile, design methodology, and customer orientation are important aspects. A framework of methods and working principles are supporting this attempt. More than 300 master-students of different disciplines and more than 250 practitioners in industry worked in small interdisciplinary teams following this TMS- approach. More than 15% of the students continued and developed a start-up quite successfully. In industry the output and the comments of participants convinced their top-management. What is the origin of this success? Just making things work and tangible in early phases of product development! Then the designers as well as the management have a much better chance to recognise the potential of ideas! This was observed and recognised during all the projects. Conclusion out of this: research regarding creativity is valuable and necessary. More important in industry—and for students—is the "making" of ideas to reduce or eliminate inadequate barriers.

U. Lindemann (✉)
Technical University Munich, Munich, Germany
e-mail: udo.lindemann@tum.de

D. Marjanović et al. (eds.), *Design Research: The Sociotechnical Aspects of Quality, Creativity, and Innovation*, https://doi.org/10.1007/978-3-031-50488-4_7

7.1 Introduction

Research in engineering design is covering a broad and often interdisciplinary spectrum. This paper is focusing on the early phases of NPD – new product development.

Specific aspects are discussed in the first part starting with involved stakeholders, requirements and vision via barriers in the process and finally addressing creativity and generating demonstrators of possible solution.

A research project as an exemplary case will be described and discussed including a summary with some general learnings out of this project.

In parallel a few aspects of designing research projects as such are discussed in this paper.

Finally, a conclusion is summarizing thoughts for designing research projects in engineering design.

7.2 A Starting Point of Product Development

7.2.1 Mission of Product Development and Expectations of Stakeholders

Product development departments in industry have to fulfill a number of different tasks like developing products based on requirements, integrating different disciplines like mechanical and electrical engineering as well as informatics, collaborate with production, sales, controlling, sub-suppliers and others, making engineering changes, managing IP-related topics, improving their competences etc.

One of the key tasks of product developers is the generation of new or improved technical products. Often, the term market offer is also used as more and more services have become an integral part of business models. Developers follow a variety of different goals, which are given or influenced by the directly or indirectly involved interest groups. The management aims to secure the company for the long term, to improve the market situation, to maximize return and to limit the risks. The shareholders expect similar goals, which, depending on the ownership structure and attitude, can lead to specific priorities. For example, in family businesses, mid- and long-term corporate development is often seen as a key element, while in capital firms' short-term returns and the performance of the stock price may dominate. Society expects secure jobs and compliance with the rules and regulations imposed by laws and regulations, taxes and duties. Customers seek to derive some kind of benefit by purchasing or using the products. And of course, the developers have their own goals, ranging from the joy and satisfaction of achieved goals and product performance to the promotion of their own career. In addition, the interests of other participants, such as in sales, production and logistics or finances are considered.

And all these stakeholders have to be taken into account during product development!

•

And the stakeholders in research? Do researchers respect the expectations of all the different stakeholders extending from researchers themselves, the scientific community via *funding organisations and publishers toward the society?*

7.2.2 Requirements

All the stakeholders mentioned above may formulate requirements based on their expectations. Quite often the contracting authority, which may be a customer or the company's management, starts already with detailed requirement lists. They may include targets with regard to cost, schedule, safety, property right, legal aspects, reliability and availability, noise, dimensions, interfaces, environmental aspects etc. All these aspects will be important during the development process, but during the early phase of a new product development process, too, many details may hinder creative and innovative solutions.

An example may demonstrate this. A customer was looking for a test rig for a high-power hydrostatic gear unit. The tender document included detailed requirements regarding the solution and insisted that the offer should not differ in any point. One potential supplier sent an offer as demanded and in addition an alternative offer based on a different type of engines. This alternative could be built with lower cost and a shorter overall length of the system. In the end, the difference in length made the customer rethink his requirements, as in this case the alternative system could be erected within an existing building instead of erecting a new one.

When developing new products, the starting point of product development should be more like a vision. But finally, all the given requirements should be fulfilled in a sufficient manner!

•

And in research? to what extend are visions instead of promised results accepted by funding organizations and reviewers?

Is it really sufficient to be rigor and demonstrate a list of well cited publications? What is the status of impact in teaching, in industry and society as a whole? Are researchers always aware that the society is financing all the research via *science foundations, industry* etc.*?*

In *addition to the research contract, the impact of research should reflect all the expectations of all stakeholders!*

7.2.3 Dynamic Changes of Requirements

The additional important aspects to be considered are the dynamic changes of technologies, attitudes and expectations of customers, management, employees and society. Since the development of market services always takes place for a time window in the future, the future situation and its boundary conditions must be anticipated. Depending on the market, time horizon and technologies used, a specific picture emerges. Thus, aircraft and rail vehicles are being developed with decades of horizons, whereas consumer electronics such as televisions or smartphones are subject to cycles of months or perhaps one year. Customers often follow these new devices directly or with some time delay of perhaps two years, whereas capital goods such as machine tools are planned for many years. In one case, the rapid development of technology in electronics and software is perceived through the replacement of the entire product; in the other, this change in technology must be carried out by exchanging subsystems, retrofitting or carrying out a fundamental and costly modernization.

In addition to technology the increasing competition by well known as well as totally new competitors, new business models, trade barriers etc. have to be managed.

To handle these dynamic changes over time a high level of flexibility is required in products as such as well as in the product development process especially during the early phases. in addition, late changes may be critical for the success of projects, as the risk of time and cost overrun is increasing dramatically over project duration.

All these challenges require the adaptability of products as well as high flexibility in all the product development processes and organizations. One of the key difficulties on the way to gain this are the mind sets of involved people.

•

In research usually there are contracts between researchers (e.g. the university) and funding institutions, and often it is not easy to change the goals (requirements) because of their impact on results in relation to finances.

In case of funding by industry partners they may ask for changes based on the actual situation in markets or the general technology development. There are also influences based on a new management structure and a modification of the company's strategy. In case of major changes this may affect the future career of researchers (e.g. PhD-students) severely.

Usual and normal causes for changes are intermediate findings differing from those expected at the beginning.

7.2.4 Human Behavior is Influencing Processes

Individual characteristics as well as the corporate culture shape the behavior of staff members in industry. The culture of dealing with failures and resulting consequences is an important indicator.

Continuing earnings pressure on the top-management leads to a risk avoidance strategy, especially for capital companies. Directors have temporary contracts and often have to deliver quarterly reports to the owners, which at least hampers long-term strategies. Controlling in companies has often focused on the topic of control and sophisticated process specifications. From the outside, the legal framework conditions have been tightened ever closer as for example in liability or the compliance issues. This also leads to avoidance of risks at company level as well as among the individual players.

Employees at the operational level are worried about their career or even their job in case of failures and the occurrence of technical and economic risks. Concerns about market or safety agencies acceptance are frequently recounted, again and again hidden behind sham arguments like leadership-weakness. Even personal envy, misunderstood competition between colleagues or the "settlement of old bills" between individual actors can play into the discussion here. Now and then ideas and suggestions are simply not understood and not admitted.

One of the key conclusions based on this discussion is a massive improvement of the effectiveness and efficiency of collaboration! Individuals as well as enterprises have to be trained and motivated to find better solutions for cooperation and collaboration, within their company as well as with external partners.

•

And in research? Collaboration especially of different disciplines often goes along with some kind of a cultural clash, misunderstanding regarding concepts and their meaning, tradition of publishing, inadequate level of trust etc.

Evaluating of research projects of other disciplines or a number of different disciplines within one project often causes misinterpretation and negative votes.

7.2.5 Barriers in and for Product Development

7.2.5.1 An Example in Industry

In a group of developers of an automobile manufacturer, the idea for a new version of a subsystem for the operation of a specific function was created, which was not the actual task of this group. After a first elaboration, a presentation on the basis of text and graphics was carried out in front of responsible managers of different corporate functions such as controlling, product safety, design, production, the actually responsible development department and others. The idea was unanimously rejected as too expensive, not approvable etc.

The developer group did not give up yet and created a physical demonstrator which was presented in a vehicle to the responsible decision makers again. The presentation took place exclusively in the concrete vehicle, as a physical object and with the required function. All of a sudden everyone was enthusiastic and supported to take it over in series production.

Summary: The team had sufficient frustration tolerance and did not give up after the first refusal. The physical realization of the new solution made a significant contribution to the fact that it was also understood and positively perceived during use.

This example indicates the importance and the influence of the management on successful innovation processes in practise.

7.2.5.2 Barriers on the Way to Decisions in Practice

A general limitation is set by the legal framework in a strict way and by the acceptance of society.

The company's history has quite an influence. Old and established companies have gone through different phases of economy, of growth and shrinking, of greater or lesser changes in organization, market orientation, good or bad economic situations. Companies run by private owners or with dominating and stable shareholders usually are focused on long term development instead of short-term figures.

This has also a large impact on the company's culture and governance philosophy. Questions of transparency, delegation of responsibility and trust are addressed as well as the entrepreneurial spirit. This also influences the way in which management and staff handle the compliance rules.

Based on that, the above discussed aspects and experience by the author a number of barriers hindering innovation may be observed:

- Risk avoidance at management level, although all innovation attempts carry some risk
- Risk avoidance by staff because of a poor failure culture to prevent negative impact regarding their career and their job

- Strict controlling of budgets and projects, hindering the staff to try out new ideas up to a certain level
- Detailed description of all processes and the monitoring of following all the linked rules in detail
- A lack of balancing chances and risks – neglecting the potential of ideas - missing professional handling of uncertainties in the sense an entrepreneurial way of thinking
- Making decisions based on insufficient information and knowledge. Quality of presentations (often done in very short time slots) has a lot of influence on decisions
- Hierarchies within organizational structures often hinder the exchange between specialists in different departments
- Hierarchies are often one of the reasons for long lasting decision processes
- Missing knowledge about customer's real need ("we know better")
- Lack of competences
- Collaboration across different departments and disciplines may cause problems due to different cultures and missions.
- Missing trust between partner companies and sub suppliersMixing up strategy and operation
- The Not-Invented-Here effect or resentments among colleagues etc.
- etc.

We are confronted with external factors like understanding of customer need, legal restrictions, competitors, property rights, availability of material, trading rules etc. And in addition, we have to concentrate on internal factors like the availability of finances and competences, the role of decision making, the lack of collaboration, and further human aspects.

To summarise the point of barriers: quite a number of all the barriers are addressing the behavior of people, their level of trust, competences and in an indirect way the point of diversity in disciplines, age, cultural background, and others. Moving the focus towards processes it is a question of level of bureaucracy vs. lean processes and the standardization vs. individualization of processes.

•

And there is quite a number of barriers in research too. In addition to all the behavioural points there are limitations in doing empirical research, which are reasoned in the limited repeatability of human work of individuals as well as teams, the difficulty of evaluating the starting point as well as the output of human work in an adequate way.

And there are evaluators and reviewers with negative votes because of misunderstandings, different understanding of good research, or further objective or subjective reasons.

Especially Interdisciplinary Research is Confronted with Difficulties to Get Funding as Well as the Acceptance of Publications by Journals or Conferences.

7.2.6 How to Overcome Barriers?

Quite a number of barriers such as legal restrictions or economic conditions in societies just have to be accepted as they are. What can developers do to overcome at least part of the other barriers?

An important element is to present the idea sufficiently clear to the decision-makers within an adequate time frame. In the above example as well as in various other projects, it has repeatedly been shown that demonstrators which are as realistic as possible can be very helpful. The two following examples show the very specific way of making demonstrators.

In order to have sufficient time for the creation of a demonstrator, in many cases so-called bootlegging-projects have been carried out on a larger or smaller scale. A systematic evaluation of such approaches is lacking due to general reluctance in industrial practice regarding publication. This is exemplified in the case of the development of the first BMW touring, where the actual procedure had been published only after a long time (www.auto-motor-und-sport.de). One employee had the first model built in his private workshop after senior management deemed such a vehicle unsuitable for the company. Some other outstanding cases have become known, such as the development of the first AUDI 100 (www.classicaudi.de). In this case, the new owner of the company had strictly forbidden the development of a new vehicle at Audi, which was ignored by the chief engineer. In other cases, the process is often passed on privately and many such projects remain unknown (see also Augsdorfer, 2008).

These are examples of bootlegging-projects, in German "U-Boot Projekte" (submarine projects). These projects are often run without notice of the management and without any budget (invisible). The author knows about some further projects of this kind in industry. In one of these projects the employees spend about one percent of the annual revenue within one year and generated the basis for one of the most important orders in history of this company. But, of course, a lot of these projects fail without any getting known to the management or the public. Trials to start a research project failed because of the lack of available information.

A former manager of IBM in Germany (Dueck, 2013), at that time responsible for innovation management, suggested to run an innovation project in the very early phase as a bootlegging-project, until sufficient information in form of a demonstrator was available, sufficient to convince the management. And he recommended to at least inform some important person in the management like the supervisor.

Running a bootlegging project is risky for those doing it with regard to their job and their career. On the other hand, the prospect of success also part of this adventure.

In order to overcome the bootlegging culture industry actually is trying out different model like the support or acquisition of start-ups, alternatively some kind of collaboration. There are some established companies allowing staff members to establish their start-up in the way of new daughter-companies. In total there are a number of activities trying to improve flexibility and collaboration capabilities.

•

In research a lot of effort (and time) is invested in preparing proposals for new research projects, in many cases with a relatively low success rate. Quite often researchers intensively cite supposed future evaluators of their proposal, sometimes just a tactical measure. In addition, researchers are following actual trends and use the wording fitting to these trends.

The most important rules are as follows:

Try again and improve the proposal!

Try to get influence on the funding policy!

7.2.7 From Vision and First Requirements to Ideas—Creativity

In research a lot of efforts have been invested to understand and support creativity and in practise there are a lot of training attempts to foster ideas.

One possibility is the search for existing solution ideas for further development. This requires investigations looking at available products, at competitors and sub-suppliers, other industries, existing patents and literature. The difficulty is finding search words and building the bridge between the given task and the wording or the abstraction level used by others. In case of a successful search it may be that an adaptation to the given requirements may be necessary, which may lead to some minor modification or a creativity-based process.

Another possibility is based on generative procedures including evaluation processes. Nowadays this is an interesting field in scientific research and up to now still restricted to relatively limited objects and systems. Future development influenced by artificial intelligence methods may help to expand the application areas of the generative part of these attempts, whereas the evaluation part still seems to be the bottleneck.

The third possibility is addressing creativity of humans, partly assisted by systematic procedures. There is a huge amount of methods available and documented in literature, trained by consultants and the range may be seen as starting with just intuition via brainstorming via biomimetics up to TIPS (TRIZ), just to mention a few examples.

Academic research is addressing these different possibilities in uneven intensities, underpinned by the important aspect of idea management including evaluation and documentation.

Beside the generative way creativity is one of the key research areas in this field of interest. There are conferences, journals and books concentrating on creativity and the support of creativity. Publications may be focused on creativity in general or on specific methods, models etc.

Between academia and industrial practise, the consultants are well established, they have at least some impact on the transfer mechanisms from academia to applications in practise, and they add quite often their own specific methods.

And looking into industrial life in practice, we usually find methods like brainstorming or derivate methods in everyday use. The quality of these applications is often quite poor and fits to the findings of Furnham (Furnham, 2000): the efficiency and effectiveness is often below the sum of individual work.

Of course, there are companies using successfully more sophisticated methods like biomimetics or TIPS (TRIZ). However, only some of the academic colleagues working on creativity have accepted these methods as being really valuable.

Looking back to the discussion of barriers in industry there seems to be no shortage of ideas as such. Maybe the quality of ideas is not sufficient in average! But it seems that a lot of the mentioned barriers stop ideas at this point of development.

7.2.8 From Idea to First Tangible Demonstrators: A Step to Overcome Barriers

In industrial design or architecture, it was and still is quite usual to build (realistic) virtual or tangible models for different purposes like studying and optimization by the specialists and to integrate other departments, users or the public.

In former times tangible models were quite usual in engineering departments too. Nowadays, especially in larger companies, building and testing prototypes is done in separate departments and these departments often are not equipped and trained to build demonstrators together with developers in the early concept phases.

The idea to integrate the capability of building and testing demonstrators in engineering design again was mentioned in literature quite often. For example, the first professor in engineering design methodology in Munich called it "Handversuch" (trial by hand). Later Rainer Bernard (Bernard, 1999) came up with the term "Innovation Lab", which was established in the same institute in the 1990's and part of the equipment were construction kits normally used by children, simple machine tools, a 3D-Printer and material to build electronic devices. Later this was extended by virtual reality. Other institutes did it in similar ways like MIT and especially Stanford, known for its Design Thinking concept.

The transfer and scaling for industrial purposes took some time. Support for the transfer is originated in new generations of 3D-Printers in combination with shops offering services in printing for everybody, the availability of cheap and standardized controllers in electronics, and finally the impact of agile concepts developed in software design.

An increasing number of services was established offering modern production facilities with established as well as new technologies. Today Maker-Spaces (e.g. https://www.maker-space.de) or similar offers help building demonstrators or very early prototypes. Usually the services are used by students of universities, established industry as well as Start-Ups, and by others, too. Users have to pay for the services via an annual abonnement, which in case of students is usually covered by sponsors.

Using this kind of facilities offers a number of possibilities to tackle several of the above-mentioned barriers by building demonstrators together in interdisciplinary teams.

7.3 Make Things Happen—an Example of Research Integrating Efforts in Teaching and Transfer into Industry

The following case (Böhmer, 2018) demonstrates some of the above-mentioned risks and obstacles when starting research and gives some hints to manage barriers. During the research of about 3.5 years duration the starting point was a new course for Master students, and at the end the transfer to industry including an extensive evaluation closed the project.

7.3.1 The Case

The development of a new concept of agile development started with the idea of teaching students to develop products in interdisciplinary teams. There were influences coming from the Design Thinking philosophy, the existence of the above mentioned Maker-Space, supporting students to generate Start-Up's based on "intelligent" products and the agile concepts in software-engineering.

Based on that, four professors and their teams of TUM from the departments mechanical engineering, computer science, electrical engineering and information technology as well as the school of business and the UnternehmerTUM GmbH using the possibilities of the Maker-Space developed a new teaching concept. Its goal is to come up with the concept of a mechatronic system under the conditions of an agile development. Students should experience and deepen cooperation across disciplines, the basics of agile approaches to physical products, the interaction of different technologies, the presentation of the results and at least ideas for a business model in interdisciplinary teams. By that the holistic view on problem solving in product development is at least sensitized. There is no specific task or vision formulated, only a set of boundary conditions of the project and processes as such.

The program is called "Think.Make.Start." (TMS) and is based on approaches of Design Thinking (rapid prototyping), SCRUM (process structure), LEAN (concentrate on value generation), the possibility of using the Maker-Space for the realization of a demonstrator as well as core elements of classical development methodology (how to do the development steps) (Böhmer, 2018).

Within two weeks the students develop their ideas and the demonstrator in teams, which should definitely include three disciplines. The requirement to build a mechatronic system (preferably with some kind of sophisticated controls/software) and essential steps of the SCRUM procedure like determined short meetings and the team-boards (ideation and project, vision and validation) are fixed. In addition, there are boards visualizing ideas, vision and validation.

Each team of about five students starts to fix their own vision for the solution to be developed. During this short phase a first team-building process may happen in parallel, if they do not know each other at that time.

The student teams start during the first two days as a maximum with discussing and finalizing their vision of the future application and the key problem to be solved.

During the main section of the project the solution is developed supported by methods and procedures of engineering design methodology. The tasks are planned on a daily basis starting with a short 15 min meeting in the morning, based on results of the day before and the tasks still open. As part of a daily routine problems and tasks etc. are documented visibly. During the day they split into sub-teams or individual work and at the end of the day they summarize the outcome of the day, fulfilled and still open tasks and occurred problems. One member of the team is looking after discipline regarding the short meeting and the documentation and another one is keeping an eye on the progress. Whenever helpful they build some kind of a prototype to learn about the kind of solution they are thinking of, or to discuss it with potential customers / users.

During the last days of their two-week project they focus on the final demonstrator, the basic idea of business opportunities and the final presentation. They have access to the Maker-Space and a small budget to buy equipment like sensors, motors etc.

On the last evening they present the results to the public in an exhibition and additionally in a short pitch.

An example may be the assembly of an elevator in new buildings with a number of floors up to skyscrapers. A guide rail for the elevator system has to be fixed over the whole height with quite narrow tolerances. The rail is delivered in smaller parts and is assembled manually in the future elevators space with help of a scaffold erected. The vision of the student team was to do this job based on a sophisticated robot system without any scaffold and manual work. After finalizing the TMS-project successfully the team (LEVARU http://levaru.de) continued, grew and received further financial support. The elevator industry is demonstrating a lot of interest.

Further successful examples are ParkHere (parking-space management www.park-here.eu), Hawa Dawa (measuring air pollution www.hawadawa.com), KEWAZO (scaffold-oriented material handling robotics www.kewazo.com) etc.

The key elements in these student projects are summarized in Fig. 7.1. Teams were formed by 4 to 6 highly motivated Master students out of different departments. They had to take over different roles like time-management etc. An adequate space allowing all the different steps within the process like design, production and test had to be available for all the teams together. Intensive collaboration within the teams helped to bring different disciplines and their specific knowledge together. In addition, all the teams shared experiences regarding the process, technology etc. The iterative way of early prototyping, testing and learning and improving the prototypes as well as early user integration based on these early prototypes helped to build market-oriented products on a conceptual basis within the given timeframe.

In addition, the management of this course had to solve the point of different departmental rules regarding credit points for students and—even worse—get a budget for this cross departmental activity—an ongoing fight except the generous support by a foundation. The main expenditures were the access to the Makerspace and the team-budgets for required prototyping material.

Figure 7.2 shows a typical situation with student teams within the working space; in the later industrial applications the situation was comparable.

After observing and analyzing about 40 student teams the transfer of this concept into industry started in parallel to additional student projects.

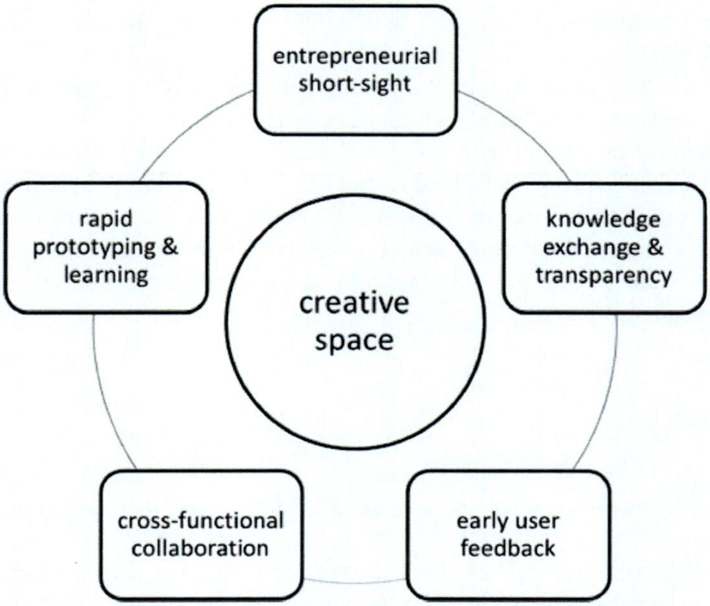

Fig. 7.1 Key-elements of think. make. start. (based on Böhmer, 2018)

Fig. 7.2 Typical TMS working situation (Technical University of Munich 2017)

Trials have been done in five companies of different size and culture, always with success. Then further projects followed in one of these companies, a large OEM. The duration was adopted to industrial need while all the other processes, methods etc. remained the same.

In total more than 200 employees participated on a voluntary basis. They spend a whole week within the same environment as the students.

The participants were very positive and noted that working together without interference by the management helped them to build demonstrators of new and innovative solutions in short time. Based on that the rest of skepticism in their management vanished. Today this company has built its own makerspace-like environment, established a specific supporting team and is working on this basis not only in the R&D departments.

7.3.2 The Summary

Looking back at risks and barriers when starting this project there are some interesting findings.

Agility is well established within the software-business, but in a broad range of companies in the mechanical or mechatronics world there were doubts regarding the potential.

Within the university just a few colleagues supported the new course—the student organisations looking after some specific budget were quite restrictive. The results were budget problems in the beginning. An external foundation liked the idea of this new student project and supported it from the very beginning. In between the top-management of the university recognized the benefits of interdisciplinary education.

Because of two weeks full time work there were some doubts to find enough interested and motivated students for the course. In between it is well established and the organizing team is selecting motivated participants based on individual interviews.

In the beginning research in this project was driven by one PhD student while within the institute some of her colleagues did not see the scientific background of her research and had some doubt regarding the project. Within the first phase of the research project there was no specific funding to finance the PhD-position available.

Later in the project a number of further researchers from different disciplines like engineering or sociology recognized the potential of observing and analysing the 18–20 student teams every year as an experimental basis.

And finally, the point of transfer and adaptation to conditions in industry was discussed. In the beginning of the project this was "just" a new course for students and some of the observers saw this as a nice "playground", but not fitting to an industrial environment. In the meantime, an increasing number of companies are using the potential of agile product development of mechatronic systems, some of them based on findings of the TMS-project.

In total it was worth to take the risks and fight against barriers. On the other hand, it was absolutely necessary to have partners in the team, recognize the growing interest in industry, and have the external support by a foundation and UnternehmerTUM.

Coming back to the title of paper, neither in student teams nor in the industry-teams a lack of creativity could be observed. The reason may be that all participants—students as well as employees in industry—did this on a voluntary basis. But this aspect was not part of the research project.

7.4 Conclusion

Designing research projects in engineering design and design researchers taking risks with regard to possible results and possibilities to publish in high ranked journals are critical points especially for younger researchers.

The scientific community should allow researchers to take more risk in their research proposals with regard to funding as well as their publications and career-relevant evaluations. Funding should be open for visions, and the evaluation should be output-oriented more than proposal-oriented. If the success-rate of research-proposals is far below 20 or 30%, this seems to be wasting of resources of researchers.

Researchers should check the actual and future relevance of a topic for industry and society, and of course do their research in a rigor and responsible way. Interdisciplinary research and trying out new methods today are linked to some kind of risk, but should not be avoided. Fascination of a research topic always has to be the motivating backbone for researchers.

Education in connection with research at least for Master students should be an integrated part of the researcher's activities.

University-management should accept that the impact of a researcher is not limited to publications and citations only and that the long-term impact of publications is not countable on a short-term basis. They should give more room for experiments in teaching and interdisciplinary research and support this.

Nowadays there are a number of changes in the engineering world bothering a number of enterprises like new business-models, product-service-systems, artificial intelligence, data analytics, additive manufacturing, individualized products and the decentralized production, sustainability and resources.

And there a number of classical research fields like creativity. In industry there seems to be enough creativity potential, but a number of barriers hindering the necessary impact of creativity. The conclusion out of that: creativity is not an important bottleneck in industry, but research is required to get a better understanding of the mechanisms of creativity.

References

Augsdorfer, P. (2008). *Managing the Unmanageable. Research-Technology Management, 51*(4), 41–47.

Bernard, R. (1999). *Early Evaluation of Product Properties within the Integrated Product Development*. Dissertation, Technical University of Munich, Munich

Böhmer, A. (2018). *When Digital Meets Physical—Agile Innovation of Mechatronic Systems.* Dissertation, Technical University of Munich, Munich.

Dueck, G. (2013). *Das Neue und seine Feinde*. Campus Verlag.

Furnham, A. (2000). The Brainstorming Myth. *Business Strategy Review, 11*(4), 21–28.

Web-sites: http://levaru.de, https://park-here.eu, www.kewazo.com, www.hawadawa.com, https://www.classicaudi.de/audi-100-allgemein/die-geschichte/, https://www.maker-space.de visited May 2019. Photo Figure 2: Andreas Heddergott, Technical University of Munich 2017.

Chapter 8
Basics of Integrated Design Engineering (IDE)

Sándor Vajna

Abstract Integrated Design Engineering (IDE) is both the further development and the enhancement of the Magdeburg model of Integrated Product Development (Burchardt, C. (2001). Ein *erweitertes Konzept für die Integrierte Produktentwicklung*). IDE is a human-centred and multidisciplinary development approach to describe and to develop products of any kind based on multi-criteria requirements from the product target groups, the humans involved in or affected by the life cycle of the product as well as of the phases of the product life. IDE thus integrates stakeholders, products and their lifecycle phases, processes, organisations, knowledge, and information. In IDE, a product is described by both its performance capability and its performance behaviour with different, but equivalent and equally important attributes in a neutral format, which offers significantly more and better ways to describe and to develop a product exactly according to various (and dynamic) requirements. IDE is based on eleven different types of integrations, at first the integration of all humans or parties involved (leading to Human Centricity), then products, processes, departments, knowledge, and methods, respectively. The underlying procedure model offers eleven activities that cover all possible development actions in arbitrary order. Dynamic navigation avoids forcing into a special work organisation and flow, thus supporting any level of flexibility within a development project.

8.1 The Product Life Cycle

IDE is not limited to specific product families, but can be used for any kind of products from any discipline (e.g. mechanics, electrics, electronics) and any kind or branch of industry. Products can be physical objects (discrete or continuous), software, methods, procedures, services or other immaterial artefacts as well as any discipline-crossing combinations thereof (e.g. mechatronic products, cyber-physical systems). This means that today's requirements to products can be met equally with solutions from different disciplines and any combinations of these.

S. Vajna (✉)
Otto-Von-Guericke-Universität, Magdeburg, Germany
e-mail: vajna@ovgu.de

D. Marjanović et al. (eds.), *Design Research: The Sociotechnical Aspects of Quality, Creativity, and Innovation*, https://doi.org/10.1007/978-3-031-50488-4_8

The majority of today's user-oriented concepts such as Human-Factor Design, Design Thinking or Human-centred software engineering focus on specific stages of the product life cycle such as development or usage. However, a holistic view on the product life cycle that encompasses all stages of a product from conceptualisation until recirculation (with the liquidation or the disposal of the leftovers as very last resort) is necessary in order to represent appropriately the complexity and interrelationships of events and facts in the real world. Many contemporary problems such as the increase of plastic in the oceans or the recycling and disposal of electronic products come from the short sightedness of focusing on specific stages of the product life cycle instead of working with a comprehensive holistic view. IDE provides such a holistic approach and anticipates complex interrelations by involving all stakeholders and stages of the product life cycle.

The product life cycle spans the phases of product planning, product development, production, acquisition and distribution, product usage (including service and maintenance), as well as recirculation. In an ideal case the product life cycle is a closed cycle, e.g. physical components of a product enter the life cycle of a new product after being refurbished or recycled. Immaterial objects such as software, methods, or procedures are replaced, re-used, or further developed and integrated into new software, methods, or procedures. Figure 8.1 provides detailed and self-explaining descriptions for each activity.

Besides the well-known distinction between product genesis and product usage, one can group the product life cycle also by the following differentiators.

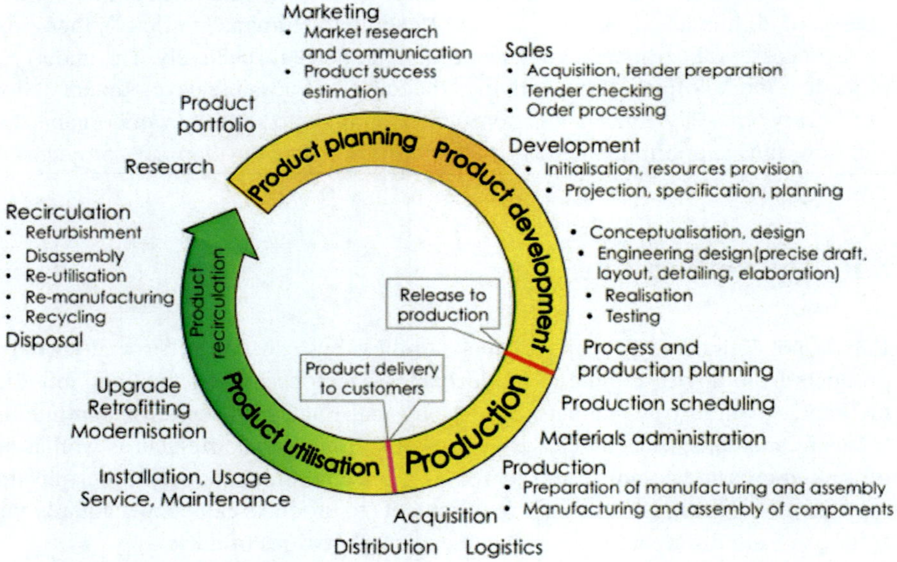

Fig. 8.1 The product life cycle and its inherent activities (Vajna, 2014)

- Virtual versus real environments: Both product planning and product development run (with the exception of preparing physical prototypes and testing them) in a virtual environment, in which the emerging product is mainly and mostly represented by and in various computer-based models. The transition between virtual environments and real/realisation environments happens at the moment of the release to production. At this transition point, all documents that describe the product, its realisation, its utilisation, and its recirculation, have to be ready and consistent. The reality is material when dealing with a physical product; it deals with implementation and test when creating software or other non-physical products (e.g. services, teaching efforts), or combinations of these, e.g. product-service-systems (PSS).
- Product ownership: The provider owns the product during its formation processes, which run until the phase of production (including distribution). After its delivery the ownership is transferred to the customer for utilisation and maintenance.
- Resources utilisation versus recirculation: After putting the product out of operation (which utilises resources) it is transferred into the recirculation phase, consisting of refurbishment, disassembly, (complete or partly) re-utilisation and recycling, and, as last resort, disposal.

Furthermore, it is necessary to look beyond the product life cycle per se, thus to consider the interrelationship of the product and the various contexts it exists within. The product life cycle is influenced by four contexts, which have (1) cultural, (2) perceptual, (3) physical, and (4) regulatory characteristics.

These contexts may affect all stages of the product life cycle as well as the humans involved in it. They may adapt to changes by e.g. general progress and by different social and political situations. In general, products interrelate in various ways with the contexts in which they are applied (Fig. 8.2).

- The cultural context emerges from cultural values, among these life style, habits, communication patterns or technological development of a culture. Cultural aspects affect the humans within the product life cycle, e.g. in the way of thinking (Nisbett, 2003), self-perception (Markus & Kitayama, 1991), and communication patterns, beliefs, and value systems. In a consequence, products have to represent the cultural values of their targeted users as well as can affect cultural elements, e.g. smart phones or text messaging have broadened the ways of communication and exchanging ideas.
- The perceptual context refers to how people perceive and comprehend the world around them. People from different cultural backgrounds perceive the world differently and process information in a different way (e.g. Nisbett, 2003). Providers and customers with different cultural backgrounds might differ in their perceptions and attitudes towards products. Cultural differences exist in regard to information processing, stereotyping, and categorization when and how products are used.

- The physical context refers to the geographical and psychological location in which a product is developed and used. The physical context affects both layout

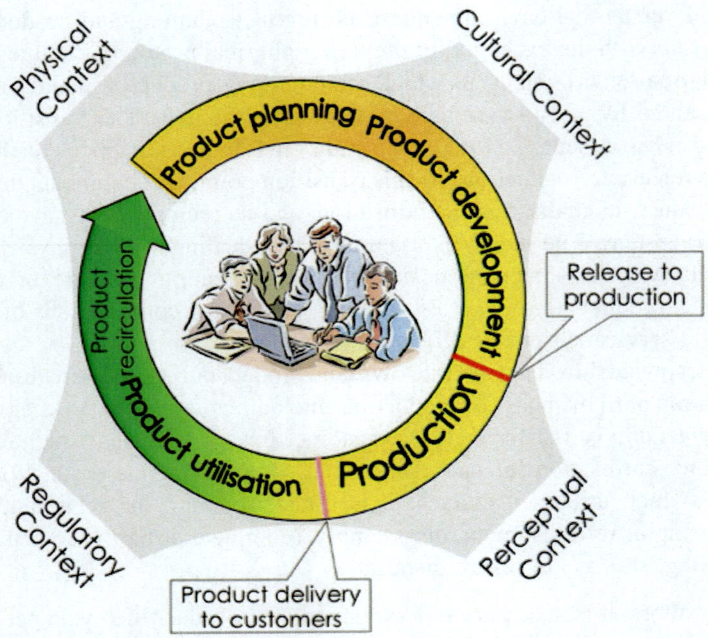

Fig. 8.2 The product life cycle with influencing contexts

and configuration of products to ensure their reliability. The physical environment and the product are mutually affected by the product application or its side effects e.g. exhaust fumes of cars causing air pollution.

- The regulatory context covers laws, regulations, standards, and guidelines. These govern the genesis, the application, and the re-utilisation of a product in its target environments on different levels of strictness. They provide a knowledge base for manifold best practices thus offering guidance when developing, manufacturing, and using solutions of any kind. On the one hand they help to minimise aberrations, on the other hand they may hinder innovations or breakthroughs, e.g. self-driving cars may require new regulations and guidelines about manufacturing issues as well as insurance policies or responsibilities in case of an accident.

8.2 Groups of Humans Involved in the Product Life Cycle

When talking of humans within the product life cycle, two groups that are involved at and directly affected by different states of the product life cycle are obvious, the customers and the providers. Customers and providers each have their own intentions, goals, and interests connected to a certain product and its realisation and utilisation processes, thus can be addressed as the stakeholders.

8.2.1 *The Customer*

The general (and main) target for setting up a product and starting its life cycle is the intention to satisfy demands. Demands are the synthesis of capabilities and limitations, needs, wishes, and (partly hidden) expectations of humans involved with the conditions for product existence in a given or changing environment (legal, social, etc.) (Ottosson, 2013). Demands come from a certain customer, who may be a single person, a group of persons, or institutions within specific markets. Each customer defines a different business, has different expectations and values, and buys something different. A customer may be a user, or a buyer, a sponsor, a patron, or combinations of these (mainly of user and buyer). The following subchapters explain briefly the intentions, goals, and interests of these groups.

8.2.1.1 The User

The user is the person or a group of persons for whom the product is aimed at and designed for, who ultimately uses or intents to use the product. The user wants a product that fulfils his specific demands effectively, efficiently, and satisfactorily. Therefore, in developing the product, special consideration have to be given also to the capabilities and limitations of the user. As mentioned above, this consideration leads to additional requirements besides those for product performance and product behaviour. The user should not be forced to adjust or to adapt to the capabilities of the product, but the product should be fitted to the user's needs, abilities, and qualifications.

Different approaches to capture the necessities of users (in the broadest sense) in order to better focus on these have been researched, among these e.g. Usability and User Experience that both focus mainly on the usage phase. For assuring this focusing on humans, the following criteria apply, which are realisation, protection from harm, building on capabilities, and personal development, Table 8.1.

Thus, the requirements for assuring human centricity add in every respect to the requirements resulting from the expected product performance. When exploring these, possible reciprocal influences have to be taken into account and balanced out. However, these effects are more than counterbalanced by focusing on humans, amongst others by much better focus on target groups, higher acceptance, and less consumption of resources during product genesis, usage, and recirculation.

8.2.1.2 The Buyer

The buyer is a person or group of persons who purchases the product. If a buyer purchases the product for his/her own use, then buyer and user are the same person. However, if the buyer is only responsible for the acquisition and the payment of the product, then buyer and user are different persons or institutions, and they pursue

Table 8.1 Criteria for focusing on humans

	Provider	User
Relisation	The provider has the necessary skills, resources, knowledge capabilities to realise the product	Basic foundation of human perception, cognitions, and motor controls are considered in the design. The user can perceive and understand information and is able to carry out operations
Protection from harm	Human working in the product life cycle are protected from impairment, dangerous, substances, in-appropriate working conditions, etc. They work in a reasonable environment Human capabilities and limitations are not overstrained	The user is protected from impairment. User's capabilities and limitations are not overstrained
Building on capabilities	Being able to apply knowledge, abilities, and skills in the work process	Building on users capabilities, knowledge, expectations, and skills
Personal development	Developing new skills, learning something new, thus gaining new knowledge and experiences In the work process	The product stimulates new experiences, helps to improve skills, and increases knowledge

different goals. The buyer is e.g. looking for low prices, for a simple and durable product without problems for many different uses, where (almost) no maintenance is required, which may lead to a conflict of interests between the buyer and the user. As a consequence, the different target groups of a product have to be clarified and to be taken into account from the very beginning of product development, even when they may contradict each other. For example, a health insurance company (the buyer) provides a wheelchair to a handicapped person (the user). The health insurance company wants a cheap, durable, and robust device that may be used by different parties. The user would appreciate a customized chair that is comfortable, not too heavy, and that can be pushed easily.

8.2.1.3 The Sponsor and the Patron

Sponsor and patron are special kinds of buyers. A sponsor links his own message to a product by using or enforcing the prestige of this product, and therefore pays the product for e.g. a user. A patron may support customers and/or providers by allocating financial, organisational, technical, or ideational means to the acquisition and/or genesis of a product. The reasons for patronage may range from philanthropy to concrete economical and/or political interests.

8.2.2 The Provider

The term "provider" is used in this context to cover all possible persons, group of persons or institutions that contribute to the realisation of a product, its transition to the consumer, and its recirculation at the end of the product life cycle. A provider could be an enterprise, manufacturer, agents, retailer, supplier, distributor, etc.

This denomination is derived from the customer's point of view. Customers get products that are, from their own view, provided from a more or less anonymous and non-transparent institution, of which the inner structure normally isn't in the focus of his interest.

- An enterprise covers all functions and activities necessary to conceptualize, to generate, and to deliver a product to a customer and to take care of the recirculation of the product at the end of its life cycle. The same is true if the enterprise has to involve suppliers because it isn't able to assure all necessary steps of product realisation
- The agent acts as an intermediary between customers and enterprises and retailers, whereby the activities of the agent covers both directions between customers and institutions.
- The manufacturer as an independent company is responsible for the production that is triggered by the input of an external ordering institution, i.e. is the supplier for this institution.
- Retailers and distributors deliver the product from suppliers to customers.
- Installers and trainers ensure user's training, product installation and product ramping up at the user's site. After-sales service assures service activities for keeping up the performance of the product.
- Re-utilisation institutions assure product re-utilisation or (as last resort) dispose the product correctly at the end of its lifetime.

8.3 Human Centricity in IDE

One central aspect of IDE is the focus on all humans within the product life cycle. This means that there are main groups of humans that are involved at different states of the product life cycle: Customers and providers (as discussed in Sect. 8.2), and, in addition, persons concerned. Taking into account the group of persons concerned as third equivalent group of humans involved, is a main difference of IDE to other development approaches.

Customers and providers have of course the strongest effect on the creation of products and are indispensable in the product life cycle, thus can be named as stakeholders. The provider ensures that an idea is turned into a product and is realised. The user customer is not only utilising the product but also ensures that the product serves its purpose and fulfils its creation objectives. However, both have to respect the necessities and expectations of the persons concerned.

Persons concerned don't have any intention and interest in dealing with a certain product as its customers and its providers do. On the contrary, creation, application, and recirculation of the product can hinder persons concerned in pursuing their own notions in their own environments and contexts. Due to their importance, persons concerned are as well stakeholders, although not with the same notions as customers and providers.

Hence, human centricity in IDE is defined as both respect for and consideration of all interests, issues, needs, and matters of *all* humans involved with a product throughout its whole life cycle. It means, as a product is generated, distributed, used, serviced, re-utilised in ways with the appropriate means and processes, that anyone dealing with the product in any of its life phases or is affected won't be overstrained in regards to his cognition, capabilities, and limitations, won't suffer any harm, and won't need to work or to live in hazardous or unethical environments, Fig. 8.3.

Human Centricity within IDE goes far beyond the various explicit and implicit considerations of the user alone, as a user-focussed view only highlights the product utilisation phase. Summarizing it can be said, that Human Centricity in IDE supports the product experience of the humans covering the complete product life cycle, i.e. it isn't only focussed on the user of the product. So, Human Centricity of in IDE covers more than the approaches of Usability and User Experience that both focus mainly on the usage phase.

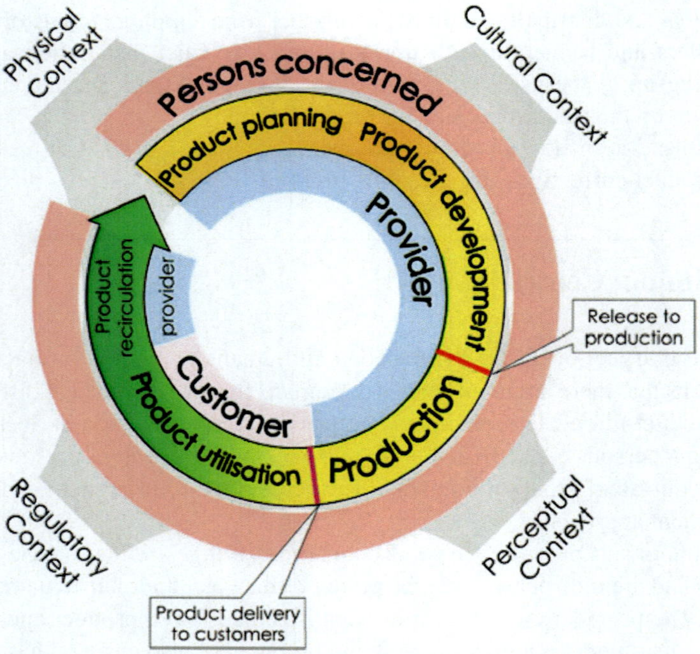

Fig. 8.3 Different groups of humans along the product life cycle within IDE

Basic ideas of human centricity have been already emphasized at the beginning of the twentieth century by Lillian M. Gilbreth and Frank B. Gilbreth, who were one of the firsts to acknowledge the crucial role of human centricity in the genesis, usage, and re-utilisation of products (Graham, 1998). They combined insights from the management of a company with findings from engineering sciences and work psychology in order to create a different view on the interdependencies between employee performance and company's working environments. In particular, they appreciated workers as individuals and did not regard them, as in other contemporary approaches, as a simple production resource, equivalent to and replaceable by machines, materials, and money.

Similarly, Shunk stated (Shunk, 1988) that the success of a company is composed of 10% hardware (i.e. factory equipment, machine tools, office equipment, computer hardware), 40% software (procedures, guidelines, software used, etc.), and 50% "peopleware", i.e. multiple-qualified employees, the so-called knowledge workers. These prefer to work in "learning companies" where work in general is regarded to be a continuous qualification (Speck & Kees, 1994). Such humans are, among others, able to react flexibly and appropriately to unforeseen disturbances or changes of customer demands.

The competency profile of humans within IDE includes professional, social, and personal skills as well as entrepreneurial skills (Hofer, 2010) (Kreis, 1997).

- Professional competence is based on the expertise that includes all knowledge, capabilities, and skills necessary to accomplish the development tasks.
- Social competence concerns all characteristics that facilitate and simplify co-existence and co-operative work in groups and teams. This also includes appropriate communication behaviour and the ability to settle conflicts without reciprocal personal harm.
- Personal competence is expressed in the correct attitude (e.g. responsibility, goal-oriented action, self-organization), in independent and holistic thinking (manifested in a timely and complete assessment of the effects of one's own actions), in the personal value system, in a realistic self-assessment, and the motivation to problem solving.
- Creative skills enable humans to form something new and valuable.
- Entrepreneurial skills encompass the will, the commitment, and the ability to directly contribute to the success of the (own) company, together with the realistic estimation of possibilities and risks of the enterprise in the market.

IDE thus creates an environment in which independent, self-responsible, active, and content people work. In this environment, the work maxim is to do the right tasks and the tasks right, both at the right time and place, instead of having to repair them later.

It has to be kept in mind that the requirements for assuring human centricity add to the requirements resulting from the expected product performance. In the following, user, provider, and persons concerned are described in detail in the view of IDE.

8.3.1 The User

The IDE approach to describe and to model the demanded (user's view) or rather the possible product performance (provider's view) by attributes fosters the accommodation of a product to a wide range of users with a large variety of capabilities and limitations.

Based on how human orientate themselves and act in their living environment, three basic needs of the user can be derived as requirements for product design that are (1) perceive, (2) comprehend, and (3) respond/operate (Fig. 8.4).

- Perceive: There are two different dimensions in perceiving that are important for product design, the aesthetic dimension and the semantic dimension (Zeh, 2010). Within the *aesthetic dimension* it is important that the product offers information that is most appropriate to the target user group to get an optimal level of stimulation in order to experience a positive affect (Zeh, 2010). In addition, this is also important to design a product in a way that it is most advanced but yet acceptable (Loewy, 1951), i.e. that there should be a well-balanced level between familiarity and newness in the design of the product (Zeh, 2010). Within the *semantic dimension* the user must be able to perceive information provided by the product. The user has to be able to find controls, perceive displayed information, feedback, and the system status. This implicit information within the product about its usage possibilities is known as Affordance (Gibson, 1977) that is part of the semantic dimension. Out of the five senses (visual, auditory, haptic, olfactory, gustatory), product design usually refers to the first three senses, visual, auditory and haptic sense, to provide information for the user. Basics of human sensation and perception such as the minimal threshold for perceiving sensory input (the minimum magnitude of a stimulus that can be discriminated from no stimulus at all) and

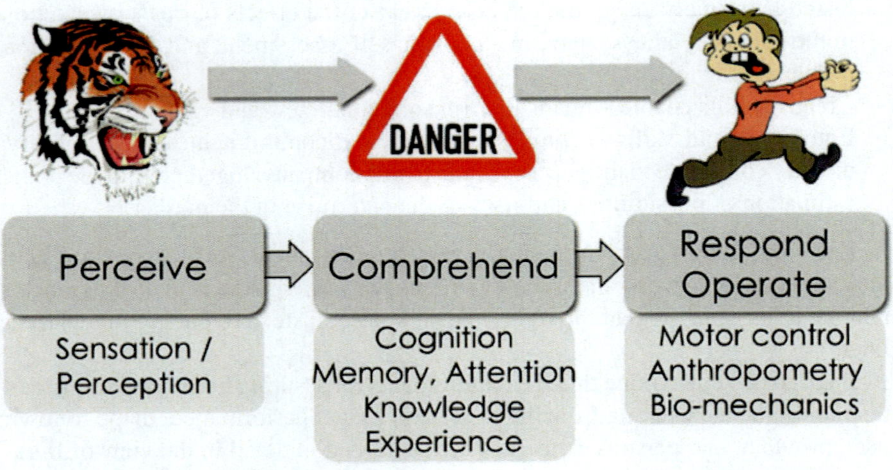

Fig. 8.4 How humans perceive, comprehend and respond to environmental information

just noticeable difference (minimum stimulus magnitude to tell two stimuli apart) have to be respected in product design.

- Comprehend: The user needs to be able to comprehend the perceived information, process it, interpret and evaluate it. Users' capabilities and limitations of cognitive processes such as memory, attention, learning, problem solving, remembering, and language have to be considered in product design and thus are mandatory in IDE. The user has to understand how to use the product. Gestalt principles have to be considered to classify information in a way that works well for human cognitive processes. Because of limitations in users' memory capacity user should not be required to hold information for a long time in working memory. Sequencing the presentation of information is preferred in order to reduce complexity. Furthermore, users' knowledge and experience have to be considered as well. Design criteria of Cognitive Ergonomics provide guidelines for increasing the comprehensibility of interfaces that are compatibility, consistency, individualization, robustness (error tolerance), transparency, learnability, memorability, efficiency and satisfaction.
- Respond/Operate: The user must be able to respond appropriately to the information provided by the product. The user must be able to operate the product safely and to carry out all functions at the expected levels. Information from the product as well as from the user's body is integrated to generate a desired motion or action. The physical dimensions of the product have to be matched to the physical dimensions (anthropometry) of the user's body to design clearance, grips, and positions of displays and controls, respectively. Capabilities and limitations of the user's motor control that is the coordination of body movements and muscle strength (biomechanics), precision of movements, and time needed to carry out specific actions have to be considered in product design.

8.3.2 The Provider

Within IDE, providers are composed of the same groups presented in Sect. 8.2.2. Some of their roles may be overlapping and the same person can be involved in different roles at different stages of the product life cycle.

Within a company, human centricity results in a change in both approach and co-operation of daily tasks, which has a positive effect on the motivation, the innovative power, and the creativity potential of humans involved. Thus, if the company acknowledges the importance of its employees, this has a beneficial effect on corporate culture and, consequently, on the motivation, innovativeness, and creativity potential of its employees.

Means of human centricity are personnel development, a motivation-oriented management, and the introduction of the continuous improvement process of products and development processes (Hofer, 2010). In IDE, the appropriate and confidence-building personnel development plays the essential role. Such a development that is both wanted by the human himself (as an intrinsic motivation) as well as required and promoted by the company (as an extrinsic motivation) results in the further development of the competence profile that leads to and fosters a human-centred, responsible, integrated, interdisciplinary, and holistic way of working in IDE.

Thus IDE is in agreement with similar ideas in modern Human Resource Management (HRM) approaches (see Aamodt for an overview (Aamodt, 2015)). Employees in an organisation are treated as key resources and not managed as factors of production. Employees are seen as crucial investment for the organisation. Fulfilling employees needs and encourage employees to develop their skills are central elements of HRM. HRM proposes that higher employee motivation and the opportunity to apply skills and abilities will improve quality and quantity of work, resulting in higher organizational productivity and profits that enable greater employee rewards and recognition, further boosting employees' motivation.

8.3.3 Persons Concerned

Persons or groups of persons concerned are part of the environments in which a product is embedded throughout its whole life cycle. Persons concerned are often overlooked in other approaches but are treated here as an important third party that needs to be considered in the product life cycle.

- The development and (mainly) the production of a product can affect persons in the physical environment of the enterprise in which the product is developed, manufactured, assembled, and serviced. The affection may range from noise annoyance by production processes, over to pollution of the environment, and end in traffic disturbances caused by transportation operations. Many companies produce their goods in 3rd world countries for the benefit of lower wages thereby causing tremendous environmental pollution.
- The utilisation of a product in a certain environment can affect or disturb directly persons in the same environment in various ways, e.g. when one is bothered by compulsory listening to musical noise created by loud music coming of the open characteristics earphones of a neighbour in a train. Another example is a leaf blower that might be beneficial for the user because it is less hard work than sweeping leafs with a broom, but that works with noise exposure and it produces exhaust fumes. Thus persons could be disturbed who have an indirect connection to the product or process.
- The recirculation of a product can disturb persons and their respective environments by various impacts when disassembling, re-assembling, recycling, and disposing the product, e.g. when toxic gas escapes during the recycling of electronic goods that threatens the people working in recycling and people living in this environment.

Although laws and regulations are trying to reduce the negative impact of certain products on affected persons by regulating product usage (like smoking prohibitions in public buildings or restaurants), product development has to strive for reducing negative impacts on affected persons as much as possible. To achieve this, the consideration of the four environments in which the product will be embedded (c.f. Fig. 8.3) is for the main part of importance throughout the development of the product. For example, catalysers in cars do not improve the performance of the car but reduce its negative impact on affected persons by polluting the air less.

8.4 Attributes

In more "traditional" design methods, the fulfilment of functions plays the main role, to which all other objectives subordinate. However, it has become apparent that the fulfilment of functions alone is no longer sufficient to meet the demands of customers (especially in the consumer goods industry). Rather, the benchmark for (long-term) customer satisfaction is the overall performance of a product (Johnson et al. 1995). In a buyer's market with many alternative products of comparable performance, the decision in favour of a product is no longer made for factual reasons alone, but increasingly for emotional reasons (e.g. spontaneous liking of a product design, "coolness" of a product), because of sustainability, simple interfaces and possibilities of use.

Within IDE, customer demands are therefore the synthesis of.

- direct customer needs, requirements, necessities, experiences, desires, open/hidden expectations, assumptions, beliefs, peer pressure,...
- market opportunities
- social, cultural, external and internal conditions, standards, constraints.

This synthesis is translated into requirements for an expected product performance (in the broadest sense), which is defined here as combination and interaction of the capabilities of a product and the possibilities to put these capabilities into utilisation (i.e. the behaviour of the product). This approach allows a description in a neutral way without pre-fixing realisation strategies and means. Product performance is modelled by eleven equal but not identical attributes resulting from the product life cycle, both from the view of the customer and from the view of the provider, considering as well the humans involved in the life cycle (provider, customer, persons affected), thus allowing a neutral product description without considering realisation strategies (Vajna, 2015).

- From the customer's view the attributes describe the customers' anticipations of the performance of the product that results from different customer needs, desires, and expectations in relation to their capabilities and limitations, as well as from the conditions of the four contexts in which the product is to be used. Attributes from the customers view form the target attribute profile.

- From the provider's view the attributes describe the performance that the product offers. Especially in the consumer goods market, this offer can also be influenced in the broadest sense by regulations and guidelines based on various customer requirements or conditions of the product environment. Attributes from the provider's view form the as-is attribute profile.

There are six product attributes that describe the behaviour and performance of the product based on customer demands.

- Product Gestalt (How does the product look like? What perceivable information is offered?) describes shape, appearance, impression and aesthetics of a product as an essential interface to the user (e.g. the design of a hair dryer, the user interface of a software product). The product developer directly determines the shape, material used and surface texture (including any coatings such as colour) as well as implicitly the structure of the product.
- Functionality (What can the product do?) describes the ability of the product to meet requirements in a measurable manner. It includes all direct and indirect functions that the product provides for use as well as their mutual influences. Functions aren't direct product traits as they arise from the interplay of certain shape, product structures, materials, and surface properties resulting from manufacturing processes.
- Usability (It the product easy to use? Is using it satisfactory?) describes in the broadest sense the performance, manageability and quality of the user interfaces of the product. It arises from the interplay of product shape, material, surface type and quality, and the existing functions.

The attributes Producibility (from the provider's point of view) and Availability (from the customer's point of view) form the two sides of the same coin. They do not complement each other, but appear alternatively, depending on whether the point of view is that of the provider (here Producibility plays the essential role) or that of the customer. Producibility and Availability are therefore treated as a single attribute.

- Producibility (Is it possible to realise the product?) provides information on whether, how, and under what conditions of a technical, an organisational and a financial nature the product can be produced either internally with the possibilities available to the provider or externally. As with the attribute Product Gestalt, the product developer directly defines manufacturing processes and thus surfaces and tolerances. A product that cannot be produced cannot be made available to the market and thus does not increase the profitability of the provider. Most customers are usually not interested in the attribute producibility, because a customer cannot and will not buy a product that cannot/could not be produced. A possible exception to this is the customer's demand for a specific (e.g. sustainable) manufacturing process.
- For the customer, Availability (Is it easy to get?) means that the product will be delivered and installed within the agreed period. For the provider, the availability of outsourced production or certain supply parts for the product can play a role.

The further common product attributes are:

- Maintainability (Is it easy to upkeep the product?) describes the ability of the product to be put back into operation, either after either having fixed deficits and failures, after having followed new requirements or changed environments, or having improved the maintenance rules. Another trigger is changing to preventive maintenance in order to avoid possible disturbances in the application area.
- Sustainability (Is the negative impact of the product on the environment low?) means that a product considers ecological aspects (e.g. renewable raw materials, renewable energies, resource efficiency) on an equal footing with technical (e.g. clean and efficient technologies, closed material cycles), social (e.g. long-term partnerships and networks of employees, suppliers, customers, competitors, administration, authorities, banks, media, etc.) and economic aspects, so that a balance can be achieved between economic, social and ecological goals (Beys, 2012).

The fulfilment of the demands are differentiated according to type, degree and quality:

- Fulfilment Type refers to kind and way a requirement is realized (for example, by selecting a suitable technical realization type from a set of alternatives that result from the provider's manufacturing capabilities).
- Fulfilment Degree describes the relationship between a requirement and its (appropriate) realization.
- Fulfilment Excellence describes both performance and value of the fulfilment.

The interaction of fulfilment type, degree, and excellence creates the so-called fulfilment level of a solution. Different fulfilment levels are described by the three fulfilment attributes Safety, Reliability and Quality.

- Safety describes the combination of the respective fulfilments of product attributes that ensure that the product does not cause any harm to the user when used as intended. This requires adequate absence or controllability of risks and hazards that may lead to failure, regardless of whether the product is used or not. Safety can be understood as a criterion for exclusion.
- Reliability is the combination of performance that ensures that a product component or the product can always be used reliably and as intended, combined with a certain tolerance to misuse (robustness) under given conditions over the specified or expected lifetime. Reliability can be understood as a minimum criterion.
- Quality describes the actually available product quality, i.e. the excellence, usability, and value of the combination of the fulfilment of the product requirements. In the use phase of the product, the quality (with given safety, reliability and availability) has a significant influence on the satisfaction of the user with this product (Ehrlenspiel & Meerkamm, 2013). Quality thus can be understood as a user and application specific list of wishes, desires, and options.

The third group of attributes covers the material and non-material aspects of the product and thus (in the broadest sense) the resulting benefits profitability. There

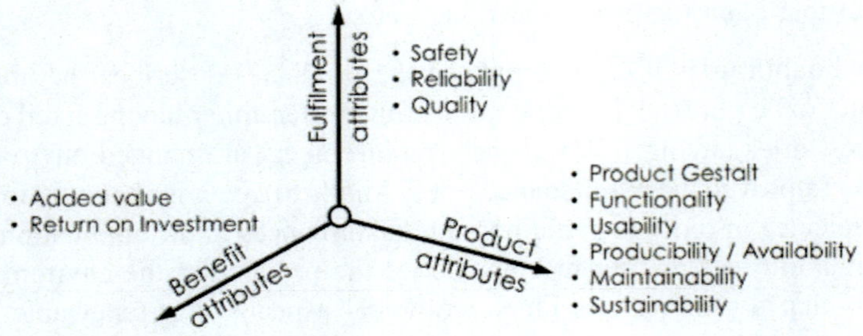

Fig. 8.5 Arrangement of the attribute groups

are two complementary attributes, because only when the supplier gets a reasonable return he will manufacture the product. Only if the customer can expect a reasonable added value from the product he will buy it.

- For the customer, added value describes not only the financial, but also the ideal increase in value through purchase, ownership, and usage of the product, i.e. when the procurement expenditure is considered to be appropriate to the achievement promise of the product. Apart from fulfilling demands, components of the gain are an additional (and unexpected) utility benefit, increased well-being, a (actual or assumed) change of state as well as a "good conscience" towards fellow human beings, the environment and/or institutions.
- Profitability is the quotient of the achievable profit from the as-is attribute profile of the product in relation to the creation effort for this profile in an accounting period. For the provider, the expected profitability is one of the main reasons to develop and manufacture the product.

Figure 8.5 shows the arrangement of the three attribute groups.

8.5 The IDE Procedure Model

The underlying holistic IDE procedure model provides, on the top level, eleven activities to create and to develop a product, which are structured in five activity groups. These activities form self-similar patterns on any level of concept, specification, and realisation of the product. Lower layers of activities provide practical approaches, procedures, methods, and tools for the activity in question, Fig. 8.6 (Vajna, 2014).

The passage usually is triggered by internal or external assignments organised as projects. However, there is no prescribed sequence in the processing, but the concrete sequence is event-driven, because it is always only determined during the current processing of the task on the basis of the actual status of the respective attribute fulfilments in context with the current status of the operational environment (Vajna, 2014). Dynamic Navigation (Freisleben, 2001) is used to provide the necessary tools

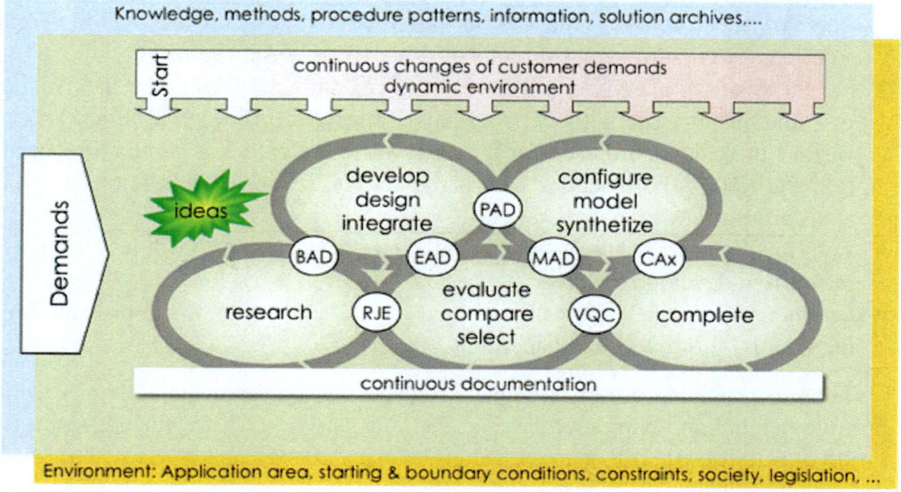

Fig. 8.6 Holistic IDE procedure model (Vajna, 2014)

and to manage the project. The task ends when providing successfully a completely developed product at the point of release to production (see Fig. 8.1).

- In order to support the easier acquisition of required information and knowledge in any activities, various *Research* activities are used, which can be carried out as rough and detailed research.
- The *Develop* activity encompasses a broad spectrum and variety of the emergence of an object, but predominantly in still small stages of concretization. The *Design* activity serves to successively work out the geometric-material totality of the object in the sense of solving the aesthetic design problem according to the product requirements formulated by the attributes in the respective environment. The *Integrate* activity merges together all emerging (partial) solutions thus assuring that demands are met even when they alter due to continuous changes of customer demands and/or dynamic environments.
- The *Model* activity includes the representation of the object in different forms of exemplifications. Such an exemplification can be a less or more complete product model (up to a digital twin), a digital mock-up according to the respective stage of development, a technical drawing and a more pictorial representation aiming at the shape, impression and surface quality of an object. The *Configure* activity is used for the concretisation and dimensioning of (emerging) objects. It does not matter whether this is a new or an adaptation development of an object or a variant construction. For an easier assessment of the object and its preliminary stages, as well as for development and interpretation, it is helpful to work with a growing computer model of the emerging object, in which the essential properties of the object, its components, its behaviour and its interaction, all of which contribute to the object's performance, can be modelled at the current stage of development.

The *Synthesise* activity balances out possible contradictions between emerging solutions by creating new solutions on a higher performance level.

- For the *Evaluation* activity of the current work status as well as for the *Comparison* and *Selection* activities of alternative objects according to arbitrary criteria, various procedures are used for the assessment, calculation, simulation, animation and testing of objects as well as the determination of economic aspects, all at arbitrary times.
- The *Completion* activity ensures that all necessary activities are done to complete the current development step.
- *Continuous documentation* of each activity takes place until the product documentation is complete and consistent.

These five activity groups are linked together by the following elements, which provide manual and computer-assisted methods and procedures for the respective group of activities (partly adapted from (Ottosson, 2013)):

- BAD (Brain-aided Design) serves to serve as an early rounding off of researched solutions. Main activities are the translation of expectations on the one hand into search terms for research, on the other hand into suitable (still abstract) solution concepts.
- PAD (Pencil-aided Design) supports fast creation, visualization, and easy fixing of solution variants as sketches ("language of the product developer") as first check of the solution feasibility.
- MAD (Model-aided Design) is used to create models of any complexity to give a first impression of shape, appearance, and dimensions of the resulting artefact ("thinking by hand"), as the evaluation of a solution can be done very easily with a model. The more complex a solution becomes, the more virtual models can be used additionally.
- EAD (Evaluation-aided Design) supports the evaluation and assessment of different intermediate and final results.
- CAx describes the use of any computer-aided system for modeling and simulating products.
- RJE (Rate, Judge, Estimate) is used to evaluate search results for usability, plausibility, consistency and coherence.
- VQC (Verify, Quantify, Check) supports evaluation and control of completion results.

Levels below the basic activity groups contain practical approaches, procedures, methods, tools and suitable CAx applications for the respective activity, as illustrated in Fig. 8.7.

Due to its self-similarity on all levels, the IDE's holistic process model can be used for modelling at the highest level of the product development process as well as at any level of detail (Vajna, 2016). It also supports the development of arbitrary objects from arbitrary domains, mechanically oriented products as well as those with an emphasis on electronics or software. The activities contained in the process model

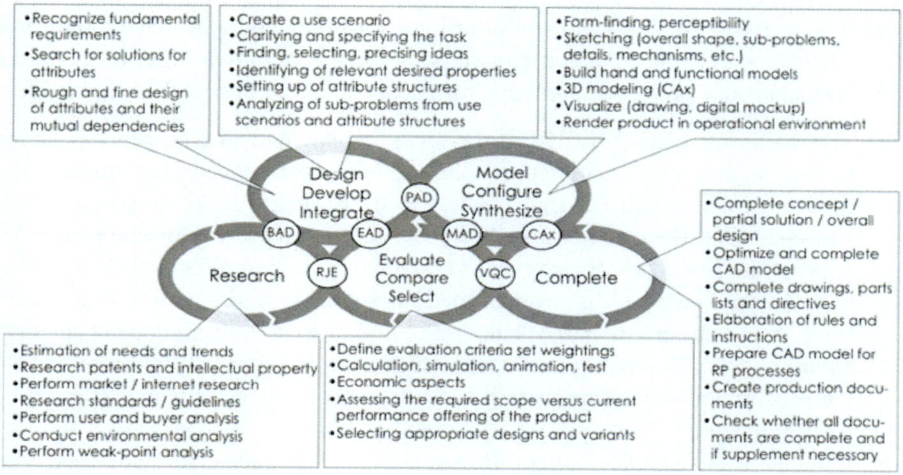

Fig. 8.7 Assigned procedures, methods, and tools in the IDE procedure model

always remain the same. The process model can therefore be used just as well to develop a concept as, for example, to develop detailed solutions for a specific product.

As an example, the IDE procedure model could be run through as follows (Neutschel, 2017): In the research activity, numerous environmental analyses (market and competitor analysis, protective right research, basic technical and design principles, etc.) take place. This allows the attribute target profile to be established and initial solution space descriptions to be developed (activity group Develop, Design, Integrate). These are compared with further research results and existing (and possibly changed) boundary conditions (activity group Evaluate, Compare, Selection), so that they are further developed into different concepts (activity group Model, Configure, Synthesize). Preferred variants are now determined (activity group Evaluate, Compare, Select), which are modelled, designed, simulated, and calculated with CAx applications (activity group Model, Configure, Synthesize). After a final comparison with existing intellectual property rights (Research activity group and Evaluate, Compare, Select activity group), the final product concept is optimized (component arrangement, material selection and thickness, etc.) and prototypically implemented in the Complete activity group.

8.6 Conclusion

Integrated Design Engineering (IDE) is a human-centred and holistic development approach that integrates products with their lifecycle phases, processes, organisations, knowledge, and information. In this paper, the four main components of IDE were described, which are Human Centricity, the applicability to any kind of products from any disciplines or combinations hereof, the neutral description of product performance and behaviour applying attributes, and the holistic procedure model.

- The main characteristic of IDE is the focus on human centricity, which is defined as both respect and consideration of all interests, issues, needs, and matters of all humans involved with a product throughout its whole life cycle. As a consequence, a product has to be generated, distributed, used, serviced, re-utilised in ways with appropriate processes. It also means that anyone dealing with the product in any of its life phases won't be overstrained in regards to his capabilities and limitations, won't suffer any harm, and won't need to work in hazardous or unethical environments. The requirements for assuring human centricity must be added to the requirements resulting from the expected product performance as described by the attributes.
- IDE enables to describe both performance capability and performance behaviour of a product in a neutral way with different but equivalent and equally important attributes. Due to this diversity IDE is not limited to a certain class of products and / or disciplines. In addition, attributes offer significantly more and better ways to describe and to develop a product exactly according to various requirements, which are based both on the needs and desires of customers as well as on the different environments in which the product is to be used and to the respective conditions of which the product must comply. The different groups (product attributes, fulfilment attributes, economic attributes) and different views (customer's view with target profile, provider's view with as-is profile, overlapping views) allow simple and easy traceable comparisons of the differences between the requirements of customers and the offer of the provider.
- The underlying holistic IDE procedure model provides eleven activities to create and to develop a product of any kind, from any disciplines, which are structured in five activity groups that cover all necessary steps to develop a product until the release to production. These activities form self-similar patterns on any level of concept, specification, and realisation of the product.

Since the year 2010, a M.Sc. course of Integrated Design Engineering has been taught at the home university of the author, in which regular semester-long IDE projects manned with 8–10 team members from four different faculties realize concrete tasks form industry, thus applying, evaluating, and further developing IDE approaches, methods, procedures, and tools, thus fostering a continuous development of IDE.

Current and future research in IDE is performed on the mapping of demands of all stakeholders (customer, provider, person concerned) to the attributes, on the measurability of fulfilments, and on the modelling of interdependencies between the attributes. One of the goals of this effort is to find out the attribute(s) in the as-is profile with which changes of demands and/or application environments can be realised at the most sustainable and the most efficient means.

References

Aamodt, M. G. (2015). *Industrial/Organizational Psychology: An Applied Approach* (8th ed.). Cengage Learning.

Beys, K. (2012), *Aachener Stiftung Kathy Beys: Lexikon der Nachhaltigkeit*. http://www.nachhaltigkeit.info/artikel/definitionen_1382.htm, accessed December 2nd, 2014.

Burchardt, C. (2001). *Ein erweitertes Konzept für die Integrierte Produktentwicklung*. Dissertation Otto-von-Guericke-Universität Magdeburg.

Ehrlenspiel, K., & Meerkamm, H. (2013). *Integrierte Produktentwicklung, fünfte überarbeitete und erweiterte Auflage*. Carl Hanser Verlag München.

Freisleben, D. (2001). *Gestaltung und Optimierung von Produktentwicklungsprozessen mit einem wissensbasierten Vorgehensmodell*. Dissertation Otto-von-Guericke-Universität Magdeburg.

Gibson, J. J. (1977). "The theory of affordances. In: Shaw, R., & Bransford, J. (eds.): *Perceiving, Acting, and Knowing*. Lawrence Erlbaum, Mahwah NJ.

Graham, L. (1998). Managing on her own: Dr. Lillian Gilbreth and women's work in the interwar area. *Engineering and Management Press Norcross*.

Hofer, M. (2010). Anwendungsgebiete von Coaching als Methode zur Förderung der Humanzentrierung in Unternehmen (Master thesis HS Mittweida 2010). http://www.alphalounge.org. Access on May 25, 2013.

Johnson, M., Anderson, E., & Fornell, C. (1995). Rational and adaptive performance expectations in a customer satisfaction framework. *Journal of Consumer Research, 21*(4), 695–707.

Kreis, B. (1977). Neue Wege zur Produktentwicklung—Untersuchung im Rahmenkonzept «Produktion 2000». *Projektträgerschaft: Fertigungstechnik und Qualitätssicherung, Forschungszentrum Karlsruhe GmbH*, Redaktion: Universität-GH Paderborn, Heinz Nixdorf Institut, Paderborn.

Loewy, R. (1951). *Never Leave Well Enough Alone*. The Johns Hopkins University Press.

Markus, H. R., & Kitayama, S. (1991). "Culture and the self: Implications for cognition, emotion, and motivation." *Psychological Review, 98*(2), 224–253.

Neutschel, B. (2017). *Parallelisierung von Produktentwicklung und Businessplangestaltung—Ein Beitrag zur Schaffung von regionalem Wachstum durch universitären Wissenstransfer*. Dissertation Otto-von-Guericke-Universität Magdeburg.

Nisbett, R. E. (2003). *The Geography of thought: How Asians and Westerners Think Differently and Why*. Free Press.

Ottosson, S. (2013). *Frontline Innovation Management* (2nd edition). Tervix Göteborg.

Shunk D (1988). "*CIM in den USA*" FB/IE 37(1988)1, pp. 19–25.

Speck, P., & Kees, U. (1994). "Auf dem Weg zum Lernunternehmen—Beispiel Werk Rohrbach der FESTO KG" Personalführung (1994) 7, S. 600–607

Vajna, S. (Ed.). (2014). *Integrated Design Engineering—Interdisziplinäres Modell für die ganzheitliche Produktentwicklung*. Springer.

Vajna, S. (2015). "Attributes in integrated design engineering—a new way to describe both performance capability and behaviour of a product". In: Weber, C., Husung, S., Cantamessa, M., Cascini, G., Marjanovic, D., & Graziosi, S. (eds.): *Proceedings of ICED15*, vol. 2, *Design Theory and Research Methodology*, pp. 2–127—2–136. Published by the Design Society, Glasgow.

Vajna, S. (2016). "The holistic procedure model of integrated design engineering". In *Proceedings of the XXX microCAD Conference 2016*, University of Miskolc (H).

Zeh, N. (2010). *Erfolgsfaktor Produktdesign*. Fördergesellschaft Produkt-Marketing Köln.

Chapter 9
Human-Centered Design Methodology as Bridge Between Academic Research and Requirements in Industry

Petra Badke-Schaub and Harald Schaub

Abstract Design Methodology provides methods and tools to assure the development of new products and services with highest quality in regard to usability, acceptance and safety. In order to make this happen it is most relevant to know about the needs and wishes of current and future users and the challenges of context variables influencing the product, the process, the economic and ecological variables. In addition to many economic criteria and best case business practices there is a "must" relate to safe and easy-to-use products. And it seems that research and practice stressing the discipline of human factors by focusing much more on safety compared to design methodology. In fact there is no commonly agreed design-for-safety methodology, however Wang and Ruxton (1998) offer a version for a design methodology of large engineering systems. Expectations and requirements in regard to safety have been defined by industry in a variety of standards and industry norms. A safe product, an easy to use product is not defined or not only defined by the respective knowledge and experience of the engineers and designers, but above all by the application and compliance with relevant standards and rules for safety, ergonomics and usability and by the acceptance of the user who buys the product (Dekker, 2006).

9.1 Introduction

The current chapter reflects in a more holistic way different results, insights but also critical questions about the contribution of design methodology and human factors in the daily design engineering work. Design Methodology in the light of the requirements of industrial and operational practice is the application of technical, structural but also psychological and physiological principles to the design of products,

P. Badke-Schaub (✉)
Technical University Delft, Delft, Netherlands
e-mail: petra.badke-schaub@gmx.de

H. Schaub
IABG & Otto-Friedrich University Bamberg, Ottobrunn, Germany
e-mail: schaub@iabg.de

D. Marjanović et al. (eds.), *Design Research: The Sociotechnical Aspects of Quality, Creativity, and Innovation*, https://doi.org/10.1007/978-3-031-50488-4_9

processes, and systems. The holistic kind of systems aims to provide an integration of knowledge and practice of human factors connecting theories, empirical research and application on a meta-level.

Whereas design methodology aims to support designers creating new design solutions which benefit the user, human factors are mainly focused on improving processes and thus by focusing on improving efficiency, creativity, productivity and job satisfaction, with the goal of minimizing errors (Brown, 1999). Designers need to have a basic understanding of principles of human factors. The goal of human factors is to reduce human error, to increase productivity, and to enhance safety and comfort with a specific focus on the interaction between the human and the thing of interest (Hollnagel, 1993). It is not simply changes or amendments to the work environment but encompasses theory, methods, data and principles all applied in the field of human factors and related disciplines (e.g. Psychology, Ergonomics, Occupational Science).

9.2 Human Factors and Design Methodology: 'one Side of Two Coins'

In literature there are few differences in terminology and definition of Human Factors and Ergonomics, what would mean that human factors, ergonomics and human-cantered design deal with the same issues of human work. The International Ergonomics Association defines ergonomics or human factors as follows: "Ergonomics (or human factors) is the scientific discipline concerned with the understanding of interactions among humans and other elements of a system, and the profession that applies theory, principles, data and methods to design in order to optimize human well-being and overall system performance." (https://www.iea.cc/about/index.html).

Many authors refer to human factors as having the same focus on the improvement of the human-product, human–machine, and human-process (in general: human-system) interface. Whereby it seems to be a characteristic of human systems to deal with new situations as combined cognitive and social tasks, including requirements and framework conditions of high complex human work (Proctor and Van Zandt, 1994). Human factors, ergonomics and human-centered design aim at the results of human work (users) and at the circumstances of the production of the products (designers). Thus human factors, ergonomics and human-centered design can be subdivided into a product perspective (users, operators) and production perspective (designers, developers).

From a product perspective the focus is often the comfortable handling, and the use of the products and the protection of personnel from physical damage. From a production perspective it is primarily to make the development process creative, efficient and effective.

Obviously, the disciplines are overlapping and cannot be completely separated, e.g. "safety" is a topic of (classical) physical ergonomics, but of course cognitive and social factors play a considerable (if not the crucial) role.

Taking the perspectives of physical, cognitive and social ergonomics the main questions can be summarized under the heading of human-centered design. Questions should be asked and scientifically investigated, e.g.

- How are decision-making processes prepared in teams with complementary, diverging or conflicting interests?
- What influence do technical aids and tools have on the decision-making process in design?
- How are information systems handled?
- What happens with Information overload?
- How does one direct the focus of attention?
- How to consolidate an integrated human, organizational and technical perspective to the product?

Human-centered design is not a new invention or another new discipline but it is a scientific approach with the highest impact on the safety of products and processes. Although the product development process integrates the safety aspect throughout the process (life-cycle design) it is still a topic which does not receive the attention it should. Often the issue safety comes too late to the surface and is often and is located in the design process only after the concept phase, mainly looking from an organisational point of view.

During the last decade Design thinking (Brown, 2019; Verganti, 2009; Badke-Schaub et al., 2005) has gained a lot of attention. One reason for the popularity of Design thinking might be due to the general recommendations, of how designers should become the accelerating engine during the product development process. One of the main contributions of Design Thinking is the anticipation of future needs and challenges of users and to implement these findings in the design of their products. The methods and results of modern human sciences support this user- and usage-centric perspective. Furthermore we claim that good design, appropriate usability and product safety can be achieved by identifying and analyzing future needs and requirements of people in their respective environments. On the basis of psychological theories integrating emotion, motivation and cognitive theories implications for design methods can be derived and implemented in hard and software.

Criteria of science are not necessarily helpful when it comes to define the main driver of successful design processes and output in industry. As Table 9.1 shows, the main important criteria of academic research have limited resemblance when it comes to their relevance in industry. For example, science requests uniqueness, but although the development of products and processes are majorly based on creativity, once the product is developed there is also a need for standardisation, rules and norms (Fig. 9.1).

Table 9.1 Design processes and requirements of academic research compared to industry

Requirements in the academic world	Requirements in the industrial world
Uniqueness	In Industry: innovation requires organizational resources
Transparency	Industry must hide new developments before they go to market
Objectivity	In industry: important is what people like and buy
Verifiability	In industry: proof in use and economic success is sufficient
Openness	In industry: openness is difficult as every mistake will be punished
Novelty	In industry: completely new solution ideas are seldom realized

9.3 Design Methodology Supports Safe Human-Centered Design

Design methods support the process of product development. Practice teaches designers and developers to focus on the form and function of the product. A safe (will not harm the user) and comfortable use (ergonomic for any user) by the everyday user often comes short. There might be several reasons for that non-fit of the product and the user however one issue makes this fit especially challenging: Usability and user experience are defined by the future user but developed and designed with the knowledge and assumptions of the designer today. The intended use of the product is determined by the designer and described in the manual and in the instructions for the product. The designer needs to foresee the behavior of the product and the behavior of the potential user in terms of use and misuse.

At the end of the day products must work, have to be safe, have to be usable and should be liked by the customer for the products´ life time. These criteria are basic for the person who wants to buy a product. Thus, the basic requirements need to be fulfilled, in order to have a chance to compete with other products. In addition to the requirements for design and function, practical experience is necessary to formulate the requirements for a safe use of the product. This is written in a large number of directives and IEC/ISO standards (e.g. IEC 61,508, ISO 26262) and determines whether and to what extent a manufacturer is liable for accidents and malfunctions of his products. From the perspective of a safe use, an exact description of the intended use of the product is necessary. This is defined in the Machinery Directive (2006/42/EC).

The Machinery Directive requires that a description of the intended use of a machine must be included in the instruction handbook. The intended use is associated with the performance of a risk assessment, which is also required by the Machinery Directive. Both technical and human factors (e.g. with regard to the necessary expertise) must be reasonably taken into account. The manufacturer of the machine must ensure that a risk assessment is carried out in order to determine the health and safety requirements applicable to the machine.

In addition to a description of the intended use, the Machinery Directive also requires consideration of the reasonably foreseeable misuse of a product. The design of the product (including the instructions for use) must not only cover the intended use of the product, but must also take into account any reasonably foreseeable misuse. Examples of reasonably foreseeable misuse include the indoor use of a grill or the use of an aggressive detergent in a food processing environment. If the manufacturer does not take into account the reasonably foreseeable improper use, this has liability consequences (Directive 85/374/EEC). However, the use and abuse result from the ingenuity of the user who can be very creative about the misuse of products, often more creative than the imagination of the designer.

In addition, human beings are also not very good in information processing. They ignore information, especially information which is not fitting within the own mental model of the situation. Human beings select information often according to the criterion of best fitting with the current situation and the current existing mental model, what is usually the most coherent and the most salient information.

9.4 Human Centered Design Approach: User-and Designer: Understanding Human Behavior

How can design methodology support the physical and above all the cognitive and social perspective in product development? Although there are various design methods and also methodologies but until now design methodologies mostly ignore cognitive and social aspects. For example, there is a human tendency to ignore information which does not fit with the individual mental model (Dörner and Schaub, 1997; Ramnarayan et al., 1994; Reason, 1990).

In industry design methodology can be applied in the area of workplace and product design (design), work support (assistant), and workplace optimization (training). In education and training, design methodology can be used in various ways to improve and support cognitive and social issues for human-factors, ergonomics or design approaches.

Design methodology (Fig. 9.1) offers several possibilities in this context. Design methodology can be used to derive the requirements for human-system interaction, educational and training situations, based on rich empiricism and theory from the field of cognitive psychology and social psychology. Which peculiarities of the human information processing system must be considered in the creation of an interaction or education/training setting, which actions, strategies, mistakes are to be expected and trained under which boundary conditions? The findings from human action regulation research are particularly useful here in order to be able to deal specifically with the cognitive and social peculiarities of the user in general and with the learner in particular (Badke-Schaub and Frankenberger, 1999).

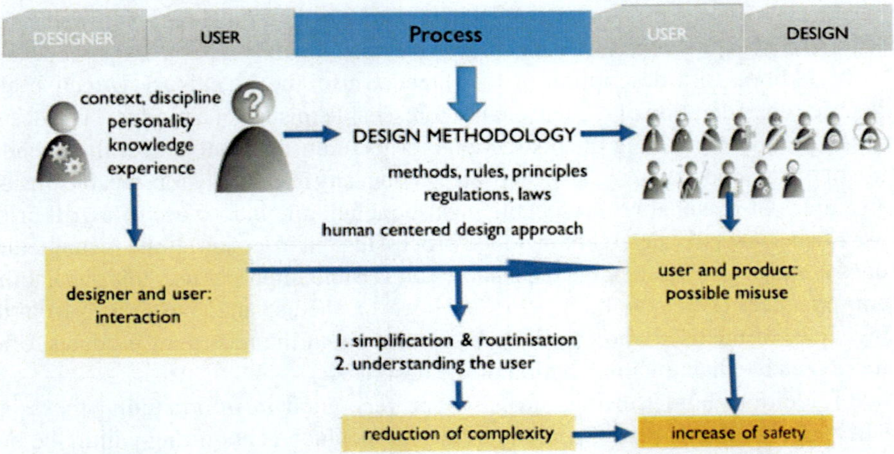

Fig. 9.1 Design Methodology as influencing support related to the topic of safety

Two forward-looking approaches of an "human enriched" design methodology are: *Simulation* of human behavior and *Work Support* with human-like assistants.

Simulations

The simulation of the expected human behavior during interaction at the workplace or with the product provides the opportunity of experimental research. With design methodology tools, artificial agents could be used to take over human characteristics at the workplace or in the interaction with a product as a surrogate. This creates the situation of a cognitive-social "dummy" acting in an experimental psycho-social wind tunnel (Cappelli et al., 2011).

Work Support

Design methodology tools can support human-like assistants who take into account the needs, abilities and shortcomings not only of the user, but also of the designer (Badke-Schaub et al., 2007; Badke-Schaub et al., 2011). The work support can be implemented in the area of communication (e.g. the assistant can suggest potential recipients of a message if he knows the way the user works and if the context situation is known), but especially in the area of information processing (e.g. the assistant can "pull" certain information from the information space taking into account the user's knowledge and typical ways of working).

The supporting human-like assistants increase the efficiency of work performance and reduce typical human errors. "Human enriched" design methodology may help in the development of these assistants.

Design methodology tools can also be part of the teaching situation that generate certain behavior with which the learner is to be trained to deal (e.g. specific cultural peculiarities can be represented, and the respective effect of a (wrong) reaction can be demonstrated to the learner).

Insights from design methodology can serve as guidance for teachers who design, select, evaluate and generate feedback in teaching situations applying knowledge of typical limitations of human information processing and communication. Using simulated human behavior fosters a human-centered design methodology.

9.5 Conclusions

Technological safety is not enough to ensure a safe product; designers need to anticipate human's thinking and acting, it needs a human-centered approach. Designers working in a human-centered way implement scientific research, and use methods and tools to improve the design process. The designer need to make prognoses about the behavior of the user under certain conditions. To do so, established methods exist such as scenario thinking, systems thinking or creativity techniques. But furthermore, designers need to forecast the behavior of the potential user in terms of use and misuse. These situations require basic psychological knowledge in the field of human factors. Obviously this situation loads an additional role onto the shoulders of the designer: the role of the psychologist. Of course, this might go too far, but the awareness of the ion Human-centered design methodology should be able to bridge the gap between scientific demands and practical requirements in industry.

References

Badke-Schaub, P. Lloyd, P, van der Lugt, R., Roozenburg, N. (2005). *Human-centered Design Methodology*. In H.H. Achten, K. Dorst, P.J. Stappers, & B. de Vries (eds.) Design Research in the Netherlands 2005, *Design Systems* Eindhoven, pp.23–32.

Badke-Schaub, P., Daalhuizen, J. & Roozenburg, N. (2011). *Towards a Designer-Centred Methodology: Descriptive Considerations and Prescriptive Reflections*. In H. Birkhofer, (ed.), The Future of design methodology. Springer.

Badke-Schaub, P. & Frankenberger, E. (1999). Analysis of design projects. *Design Studies, 20*, 481–494.

Brown, S.J. (1999). Human factors and safety integrity—IEC 61508. 156–161. https://doi.org/10.1049/cp:19990180.

Brown, T. (2019): Change by design. Design Driven Innovation: Changing the Rules of Competition by Radically Innovating What Things Mean.

Cappelli,M., Gadomski, A.M. & Sepielli, M. (2011). Human factors in nuclear power plant safety management: A socio-cognitive modeling approach using TOGA meta-theory. In *Proceedings of International Congress on Advances in Nuclear Power Plants*. Nice (FR)., SFEN, S.W.A. (2006).

Dekker, S. W. A. (2006). *The field guide to understanding human error*. Ashgate Publishing, Ltd.

Directive 2006/42/EC of the European Parliament and of the Council of 17 May 2006 on machinery, and amending Directive 95/16/EC. http://data.europa.eu/eli/dir/2006/42/oj22.02.2022

Dörner, D., & Schaub, H. (1994). Errors in planning and decision-making and the nature of human information processing. *Applied Psychology, 43*(4), 433–453.

Hollnagel, E. (1993). *Human reliability analysis: Context and control*. Academic Press.

Proctor, R. W., & Zandt, T. (1994). *Human factors in simple and complex systems*. Allyn & Bacon.

Ramnarayan, S., Strohschneider, S., & Schaub, H. (1997). Trappings of expertise and the pursuit of failure 1997. *Simulation and Gaming, 28*(1), 28.

Reason, J. (1990). *Human error*. Cambridge University Press.
Verganti, R. (2009). Design Driven Innovation: Changing the Rules of Competition by Radically Innovating. *Harvard Business School Publishing Corporation.*
Wang, J. & Ruxton, T.A. (1998). Design-for-safety. Methodology for large engineering systems. *Journal of Engineering Design, 9*(2).

Chapter 10
What Does It Mean: "Quality of Design Research"?

Christian Weber

Abstract ISO 9000:2015 defines the term "quality" as the "degree to which a set of inherent characteristics of an object fulfils requirements". The same standard also defines basic concepts for the subordinated terms like "inherent characteristics", "object", "requirements", etc. This article asks what all this means for "engineering design research" as one of the major objects in focus of this publication:

- How to decompose the object "design research" further?
- What are its (inherent) characteristics?
- What are the requirements?
- Who poses requirements?

10.1 Introduction

According to (ISO 9000:2015) the term "quality" is defined as:

Quality = Degree to which a set of inherent characteristics of an object fulfills requirements

This definition contains some more terms that are also defined in (ISO 9000:2015) or explained in attached documents. Table 10.1 shows these definitions/explanations and states the focus of this article. The remaining core issues "design research" (in the meaning of "research on design***ing***"), "characteristics" of design research and "requirements" of design research are discussed in the subsequent Sects. 10.2, 10.3 and 10.4.

This article does not provide "yet another" view on design research. Instead, it is to be understood as a material collection as a contribution to "action to identify opportunities for collaboration and consolidation" (McMahon, 2012).

C. Weber (✉)
Technische Universität Ilmenau, Ilmenau, Germany
e-mail: christian.weber@tu-ilmenau.de

D. Marjanović et al. (eds.), *Design Research: The Sociotechnical Aspects of Quality, Creativity, and Innovation*, https://doi.org/10.1007/978-3-031-50488-4_10

Table 10.1 Definition/explanation of terms in the "quality" definition according to [ISO 9000:2015]

Term	Definition/explanation	In our field/in this article	Section
Object	Anything perceivable or conceivable; examples: product, service, process, person, organization, system, resource	The object is "design research" To be further decomposed into:	
		• Research on design*s* (solutions, products) → Not investigated here	—
		• Research on design***ing*** (processes, methods, tools) → Focus here	10.2
Characteristics	Distinguishing features	What are "distinguishing features" of design research?	10.3
Inherent	As opposed to "assigned", "inherent" means existing in the object	In quality management there is big debate whether it is at all possible to distinguish between "inherent" and "assigned" characteristics? → Not investigated here	—
Requirements	Need or expectation that is stated, generally implied or obligatory … generated by different interested parties or by the organization itself	Which expectations? Stated by whom?	10.4

10.2 The Object: Design Research

10.2.1 Research in General

There is an unmanageable number of definitions for the term "research" (> 1 billion hits googling "definition research").

Relatively straightforward (and well-known) general definitions are:

- "The systematic investigation into and study of materials and sources in order to establish facts and reach new conclusions." (Oxford Dictionaries, 2018)
- "Studious inquiry or examination, especially: investigation or experimentation aimed at the discovery and interpretation of facts, revision of accepted theories or laws in the light of new facts, or practical application of such new or revised theories or laws." (Merriam-Webster, 2018)

Two more definitions application aspects of research, as is also relevant for design research:

- "Research and experimental development (R&D) comprise creative and systematic work undertaken in order to increase the stock of knowledge — including knowledge of humankind, culture and society — and to devise new applications of available knowledge." (OECD, 2015)
- "A careful, systematic, patient study and investigation in some field of knowledge, undertaken to establish facts or principles. … Research is a structured inquiry

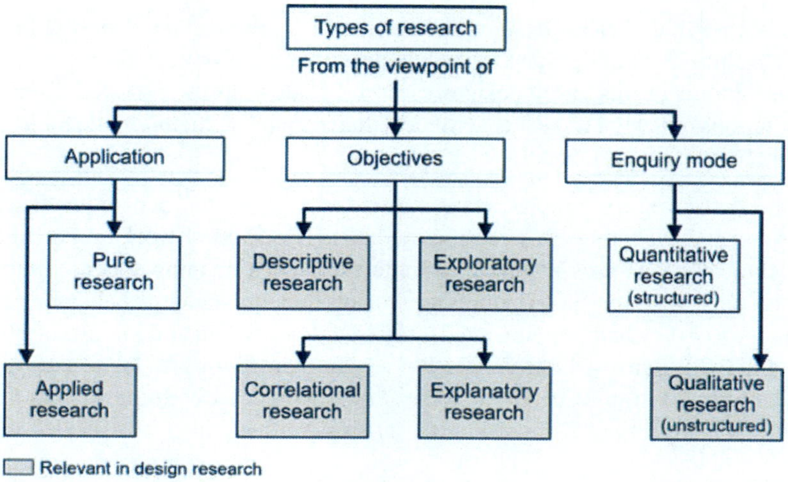

Fig. 10.1 Types of research according to (Kumar, 2011); highlighting relevant areas for design research by the author of this article

that utilises acceptable scientific methodology to solve problems and creates new knowledge that is generally applicable." (Grinell, 1993).

Kumar, 2011 from social sciences — that can deliver valuable input for design research in many respects — presents an interesting typology of research (Fig. 10.1).

Some explanations of the terms used in Fig. 10.1 (Kumar, 2011):

- "A research study classified as a ***descriptive study*** attempts to describe systematically a situation, problem, phenomenon, service or programme ... The main purpose of such studies is to describe what is prevalent with respect to the issue/problem under study.
- The main emphasis in a ***correlational study*** is to discover or establish the existence of a relationship/association/interdependence between two or more aspects of a situation. ...
- ***Explanatory research*** attempts to clarify why and how there is a relationship between two aspects of a situation or phenomenon. ...
- The fourth type of research, from the viewpoint of the objectives of a study, is called ***exploratory research***. This is when a study is undertaken with the objective either to explore an area where little is known or to investigate the possibilities of undertaking a particular research study. ... Exploratory studies are also conducted to develop, refine and/or test measurement tools and procedures."

10.2.2 Design(ing)

In "design research" the term "design" may denominate two different things:

- Designs: In this case "design research" looks at the results of design processes, i.e. solutions of design tasks, product descriptions, products, etc.
- Designing: In this case "design research" deals with the process of developing solutions of design tasks, i.e. activities, their sequence, methods and tools, societal and human aspects, etc.

The first category aims at analysing and improving the behaviour of technical elements and systems with scientific methods. This kind of work marks the origins of technical sciences as a whole, in some cases even creating specific branches of science itself (e.g. thermodynamics as a means to understand the behaviour of steam engines ***after*** they had been invented), and is still prominent and important. However, research on designs will ***not*** considered any further in this article.

The term "design research" is today more associated with the second category, aiming at analysing and improving the process of designing. This is also the focus of this article.

There is again a very big number of definitions of the term "design" (close to 1 billion hits in Google). To start with more general ones—concentrating at the word's meaning of design***ing***:

- "The art or action of conceiving of and producing a plan or drawing of something before it is made." (Oxford Dictionaries, 2018).
- "The creative art of executing aesthetic or functional designs." (Merriam-Webster, 2018).

Some relevant (and well-known) definitions of the field of design research itself are:

- "Engineering Design is the use of scientific principles, technical information, and imagination in the definition of a product, machine or system to perform prescribed functions with the maximum economy and efficiency." (Feilden, 1963)
- "The totality of the activities with which all the information necessary for producing and operating a technical system or product is processed in accordance with the task. The result is a set of product documents." (VDI 2221:1987)

Maybe one of the shortest, however far-reaching definitions:

- "The conception and planning of the artificial." (Simon, 1969)

10.2.3 Design Research

A first approach to define "design research" is to combine definitions of "research" with those of "design". An example:

- A careful, systematic, patient study and investigation ***of the conception and planning of the artificial***, undertaken to establish facts or principles. ... Research is a structured inquiry that utilises acceptable scientific methodology to solve problems ***of the conception and planning of the artificial*** and creates new knowledge that is generally applicable (Grinell, 1993) combined with (Simon, 1969).

Again, when searching for more specific definitions, we find a multitude of approaches (about 1 billion hits when googling "design research"). Among other things, the question still inspires interesting scientific debates, like the one sparked off by Imre Horváth on Researchgate (Horváth, 2018). Here, as in many cases, there seems to be a fluent transition between the terms "design research" and "design science".

A selection of statements of well-known authors is:

- "Design science is a system of logically related knowledge that should contain and organise the complete knowledge about object and process, and about and for designing." (Hubka and Eder, 1996).
- "Design research aims at increasing our understanding of the phenomena of design in all its complexity, and at the development and validation of knowledge, methods, and tools to improve the current situation in design." (Blessing, 2002).
- "Design science studies the phenomena of design above the disciplinary manifestations and achieves an integral view by applying both aggregation and abstraction." (Horváth, 2004).
- "Design science studies the creation of artifacts and their embedding in our physical, psychological, economic, social and virtual environments." (Papalambros 2015).

Horváth, 2004 also proposes a detailed taxonomy of design research which is very interesting reading but, for reasons of space, not discussed here.

In the annual Summer School on Engineering Design Research (SSEDR) — the aims and results of design research are characterised as follows (Blessing 2018):

- "Aims [of design research]:
  - The formulation and validation of models and theories about the phenomenon of design, as well as
  - the development and validation of design support[1] — founded on these models and theories — to improve design(ing)

- Results [of design research] are contributions to:
  - Knowledge (empirical findings),
  - design theory (empirical findings, organisation and reasoning),
  - work practice and education (support and artefacts/services)".

---

[1] Support: The possible means, aids and measures that can be used to improve design, incl. strategies, methodologies, procedures, methods, techniques, software, knowledge, guidelines, information sources, training, etc., addressing one or more aspects of design.

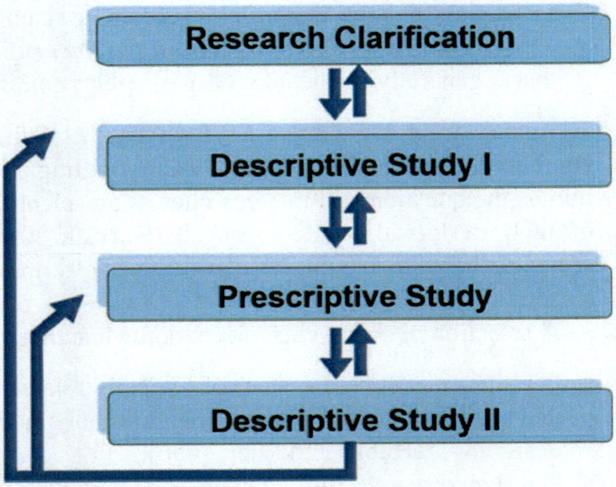

Fig. 10.2 Stages of the design research methodology (DRM) (Blessing and Chakrabarti, 2009)

It is evident from practically all definitions and approaches that in design research — different from many other scientific disciplines — descriptive research (analysis) is not enough; design research also aims at improving the current situation (prescriptive research, synthesis). This is strongly reflected in the Design Research Methodology (DRM) proposed by (Blessing and Chakrabarti, 2009) — so far the only comprehensive concept of how to do and systematise design research activities (scheme in Fig. 10.2). The Design Research Methodology is, therefore, the base of instructing young design researchers in the framework of the annual Summer School of Engineering Design Research (SSEDR).

10.3 Characteristics of Design Research

10.3.1 Characteristics of Research in General

Again, we find an unmanageable number of definitions for the term "characteristics of research" in general (> 500 million hits in Google). Again coming back to social sciences, (Kumar, 2011) states that (good) research has to have the following characteristics:

- "***Controlled:*** In real life there are many factors that affect an outcome. A particular event is seldom the result of a one-to-one relationship. Some relationships are more complex than others. Most outcomes are a sequel to the interplay of a multiplicity of relationships and interacting factors. In a study of cause-and-effect relationships it is important to be able to link the effect(s) with the cause(s) and vice versa. In the study of causation, the establishment of this linkage is essential; however, in practice, particularly in the social sciences, it is extremely difficult — and often impossible — to make the link. ...

- ***Rigorous:*** You must be scrupulous in ensuring that the procedures followed to find answers to questions are relevant, appropriate and justified. Again, the degree of rigour varies markedly between the physical and the social sciences and within the social sciences.
- ***Systematic:*** This implies that the procedures adopted to undertake an investigation follow a certain logical sequence. The different steps cannot be taken in a haphazard way. Some procedures must follow others.
- ***Valid and verifiable:*** This concept implies that whatever you conclude on the basis of your findings is correct and can be verified by you and others.
- ***Empirical:*** This means that any conclusions drawn are based upon hard evidence gathered from information collected from real-life experiences or observations.
- ***Critical:*** Critical scrutiny of the procedures used and the methods employed is crucial to a research enquiry. The process of investigation must be foolproof and free from any drawbacks. The process adopted and the procedures used must be able to withstand critical scrutiny."

Guide2Research, 2014, a research portal for computer science (meanwhile transferred into Research.com), distinguishes between good and bad research:

- "The main characteristics for good quality research is listed below:
  - It is based on the work of others.
  - It can be replicated and is doable.
  - It is generalisable to other settings.
  - It is based on some logical rationale and tied to theory. In a way that it has the potential to suggest directions for future research.
  - It generates new questions or is cyclical in nature.
  - It is incremental.
  - It addresses directly or indirectly some real problem in the world.
  - It clearly states the variables or constructs to be examined.
  - Valid and verifiable such that whatever you conclude on the basis of your findings is correct and can be verified by you and others.
  - The researcher is sincerely interested and/or invested in this research.
- Meanwhile, bad research has the following properties:
  - The opposites of what has been discussed.
  - Looking for something when it simply is not to be found.
  - Plagiarizing other people's work.
  - Falsifying data to prove a point.
  - Misrepresenting information and misleading participants."

It is interesting to note that replicability is included in the second list of characteristics of (good) research (according to [Guide2Research, 2014]), but is not part of the first list (according to (Kumar, 2011)). The author's explanation is that the first approach comes from social sciences. Here, like in design research or design science, respectively, it is nearly impossible to repeat experiments: One cannot replicate the same conditions (e.g. the same [set of] people, having the same knowledge, working on the same task in two empirical design studies at different times, e.g. testing the effectiveness of a method or tool). This may also explain why many concepts of research from social sciences fit well to the situation in design research/science.

A way out could be virtual designers and design groups, represented by software agents. This, however, is today more a vision than a practical proposition.

10.3.2 Characteristics of Design Research

If we look closer into the area in question here (design research), there are relatively few concepts of what is good design research. In the Summer School on Engineering Design Research (SSEDR) the following hints are given (Blessing, 2018), most of them based on (Cross, 1999):

- Purposive:
  - Based on identification of an issue or problem worthy and capable of investigation.
- Inquisitive:
  - Seeking to acquire new knowledge.
  - Informed, conducted from an awareness of previous, related research.
- Methodical:
  - Planned and carried out in a disciplined manner.
- Communicable:
  - Generating and reporting results which are testable and accessible by others.
- Non-discriminatory:
  - Independent and free from any direct or indirect censorship.
- Influential (M.M. Andreasen):
  - Change the way we (researchers and designers) think and act.

When A. Chakrabarti and L. T. M. Blessing edited their book on an Anthology of Theories and Models of Design (Chakrabarti and Blessing, 2014) all contributors were asked to present their views of design research/design science. The author of this article noted down the following characteristics (Weber, 2014):

- A model or theory of designs and designing, like any scientific statement or theory, must explain and predict observations in its field.
- Since (Popper, 1935) or (Popper, 1959), respectively, we want a model or theory falsifiable rather than verifiable.

- A model or theory of designs and designing should have two sides:
  - Collecting and systematising knowledge about "what is" (descriptive part).
  - Collecting and systematising knowledge about actions and skills that can change the present state into another, previously not existing state (prescriptive).
- Any science deals with certain objects in its field. In the case of designs and designing there are two different "objects" to be considered:
  - The designs (as artefacts)
  - The designing (as a rationally captured process to create artefacts).
- A model or theory of designs and designing is "situated" in the sense that external influences (knowledge in other fields, society, markets, new technologies, time, ...) have to be considered as they evolve, resulting in modified or new models and theories.
- There may be different "stakeholders" who pose requirements (maybe: "demands") on models and theories of designs and designing. ... (See next section)
- We may state that an "appropriate model or theory of designs and designing", beyond the usual criteria of a descriptive science (e.g. truth, completeness, level of detail), has to meet criteria like "usefulness" (for different stakeholders!) and "timeliness".

10.4 Requirements of Design Research

Weber and Birkhofer 2007 argued that it is perfectly acceptable to pose requirements on any type of science in general, but on design research/design science in particular. This view is also perfectly in line with the quality definition according to [ISO 9000:2015] (see Sect. 10.1).

In (Weber and Birkhofer, 2007) the following groups of requirements of design research/design science were identified:

- Establishing and systematising the knowledge in the field (including terminology)
- Guidelines for scientific reasoning
- Coherent descriptions and prescriptions of products (designs)
- Coherent description and prescription of design processes (designing)
- "Map" of knowledge (product-/process-related as well as descriptive/prescriptive)

The next question is: Who may pose these requirements?

In (Weber and Birkhofer, 2007) the following stakeholders (groups of "interested parties" in the diction of [ISO 9000:2015]) were proposed:

- Scientists
- Designers in practice
- Students (including PhD students/candidates)
- Tool/software developers

The considerations described in (Weber and Birkhofer, 2007) ended in a table that related requirements of design research/design science to stakeholders—here reproduced as Table 10.2.

The groups of stakeholder posing requirements on design research/design science, the groups of requirements and their further decomposition, the relations between the two according to Table 10.2: All this could and should be discussed in the community. However, the proposals decribed in (Weber and Birkhofer 2007) and, in part, repeated here could be a start from which to proceed.

10.5 Conclusions and Challenges

This article tries to apply concepts of (ISO 9000:2015) to the "quality of design research" and it investigates important terms. These must be filled with our own content. The main challenges are:

- Clarify relations between basic terms like "design science", "design research" and maybe "design knowledge"; define their relations to "design practice" and "design teaching".

A first proposal is given in Fig. 10.3. The relations marked (a) and (b) are quite special and have to be discussed further:

(a) Do we accept research that only contributes to practice but not to the acquisition of general knowledge as "science"?
(b) Quite often new or revised processes and methods are not tested with practitioners but with student groups. How valid are the results? (See also Design debate at the DESIGN 2018 conference.)

- Define characteristics of (good) design research.
- Consider that descriptive research (analysis) is not enough, we want/have to improve things (prescriptive research, synthesis)—first proposal in (Blessing and Chakrabarti, 2009), see Fig. 10.2.
- "Usefulness" as an additional criterion?!
- Generalisability/verification/repeatability problematic in our field (like in social sciences!).
- Consider multi-domain concepts of designs and designing.
- Consider the dynamics of the research object(s): designs, designing.
- Who are the stakeholders of design research ("interested parties") and what are their requirements?

We have still some work to do—fortunately!

Table 10.2 Requirements of design research/design science related to stakeholders

Requirement	Scientists	Practitioners	Students			Tool-/ SW-developers
			Undergrad	Graduate	PhD	
Establishing and systematising terminology	●		○	○	●	○
Scientific reasoning:						
– Formulation of hypotheses, findings, conclusions	●				●	
– Research methodology	●				●	
– Verification/ falsification methods & tools	●	○			○	?
Coherent descriptions and prescriptions of products:						
– Properties and their relations	●	●	●	●	●	●
– Multitude of properties and relations	●	●	○	○	○	●
– Multi-discipline/ -domain approach	●	●	○	○	○	●
– Formalisation	●	○	○	○	●	●
– Modularisation		●	○	○	○	●
– Development/ application of tools for product modelling	○	●		○	○	●
– Acceptance of methods and tools	●	●			●	●
– Relation to business goals		●	○	○	●	●
– Integration into existing environments		●			○	●
Coherent descriptions and prescriptions of design processes:						

(continued)

Table 10.2 (continued)

Requirement	Scientists	Practitioners	Students			Tool-/ SW-developers
			Undergrad	Graduate	PhD	
– General framework	●	○	•	•	●	●
– Specific/ "situated " processes	●	●	○	○	●	●
– Assigning methods and tools to processes	●	●		○	●	●
– Assessing methods and tools	○	•		○	●	?
– Formalisation	●	○	○	○	●	●
– Modularisation, designer's workbench		●		○	●	●
– Devel./ application of tools for process support	○	●		○	○	●
– Acceptance of methods and tools	●	●			●	●
– Work distribution	○	●		○	○	●
– Integration into existing environments		●			○	●
"Map" of product- and process-related as well as descriptive/ prescriptive knowledge	●	○	○	○	●	○

● Core issue; ○ Depending on particular project

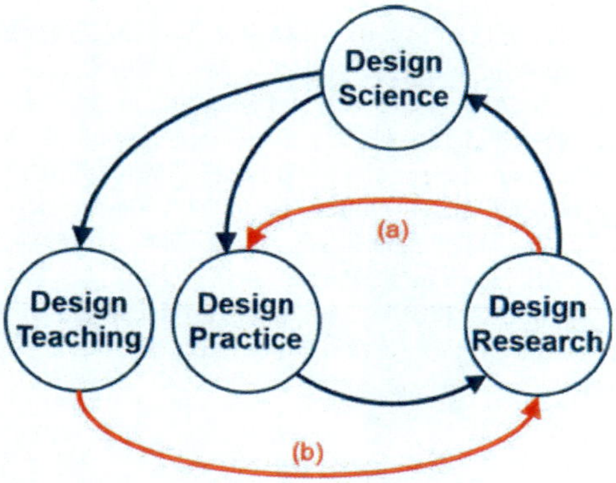

Fig. 10.3 A proposal for the relations of basic terms in design research / design science (see additional explanations in the text)

References

Blessing, L. T. M. (2002). What is this thing called design research? *Annals of the 2002 international CIRP design seminar*, 1–6.

Blessing, L. T. M., & Chakrabarti, A. (2009). *DRM—a Design Research Methodology*. Springer.

Blessing, L. T. M. (2018). Introduction to design research. In *20th Summer School on Engineering Design Research* (SSEDR).

Chakrabarti, A., & Blessing, L. T. M. (eds.) (2014). *An Anthology of Theories and Models of Design*. Springer.

Cross, N. (1999). Design research — a disciplined conversation. *Design Issues, 15*(2), pp. 5–10.

Oxford Dictionaries. https://en.oxforddictionaries.com/definition/research.

Feilden, G. B. R. (1963). *Engineering design. Report of a Committee appointed by the Council for Scientific and Industrial Research ("the Feilden Report")*.

Grinell, R. M. (1993). *Social work research and evaluation*. Peacock Publishers.

Guide2Research. (2014). *Research Blog, Tutorials: Top 10 Qualities of Good Academic Research*, posted on February 12, 2014. https://research.com/research/top-10-qualities-of-good-academic-research

Horváth, I. (2004). A treatise on order in engineering design research. *Research in Engineering Design, 15*, 155–181.

Horváth, I. (2018). Posing the question: "What is design research? *What is design science? More definitions, than authors ... Can we come to a general definition or it has no sense to strive for it?"* https://www.researchgate.net/post/What_is_design_research_What_is_design_science.

Hubka, V., & Eder, W. E. (1996). *Design Science*. Springer.

ISO 9000:2015 (2015). *Quality management systems — fundamentals and vocabulary*. International Standardization Organisation, Technical Committee ISO/TC 176/SC 1 „Concepts and terminology".

Kumar, R. (2011). *Research Methodology* (3rd ed.). Sage.

McMahon, C.A. (2012). Reflections on diversity in design research. *Journal of Engineering Design, 23*(8), 563–576. https://doi.org/10.1080/09544828.2012.676634

Merriam-Webster (2018). https://www.merriam-webster.com/dictionary/research

OECD (2015). *Organisation for Economic Co-operation and Development (OECD): Frascati Manual 2015: Guidelines for Collecting and Reporting Data on Research and Experimental Development* (p. 2015). OECD Publishing.

Papalambros, P.Y. (2015) (with contributions by 28 additional researchers): Design science: Why, What and How? *Editorial of the inaugural volume of the Design Science Journal*.

Popper, K. R. (1935). *Logik der Forschung*. Julius Springer.

Popper, K. R. (1959). *Logic of Scientific Discovery*. Hutchinson & Co.

Simon, H.A. (1969). *The Sciences of the Artificial.* MIT Press.

VDI-Guideline 2221 (1986/1993). *Methodik zum Entwickeln und Konstruieren technischer Systeme und Produkte.* VDI, 1st ed. 1986, updated version 1993.

VDI-Guideline (1987). *Systematic Approach to the Design of Technical Systems and Products.* English version of VDI 2221 (1986/1993).

Weber, C., & Birkhofer, H. (2007). Today's requirements on engineering design science. In *Proceedings of ICED 07 (DS 42)*, pp. 785–786 (Exec. Summ.), Paper-no. 511 (Full Paper, CD-ROM), École Centrale Paris.

Weber, C. (2014). Modelling Products and Product Development Based on Characteristics and Properties. In: (Chakrabarti & Blessing 2014), chapter 16, pp. 327–352

Chapter 11
Conclusion

Dorian Marjanović, Mario Štorga, and Stanko Škec

The workshop "The Design Research—Sociotechnical Aspects of Quality, Creativity and Innovation" provided a framework for a comprehensive exploration of various aspects of design research, education, and practice. Discussion and presentations at the workshop trace the evolution of design research over the past decades, highlighting its growth, consolidation, and diversification. Each contribution provides the author's view on the multidimensional nature of design, incorporating technical, social, and policy attributes. The challenges in front of design research are huge. The requirements of all stakeholders, industry, users, policymakers and society are constantly changing. Practitioners and researchers must cope with such a need. The content presented does not intend to trace future research. Instead, 20 reputable authors present their experience, vision and competence. The views presented are diverse. Nevertheless, all the presentations and workshop discussions indicate the importance of all three aspects of design research considered: Quality, Creativity and Innovation.

The historical development of design research is briefly presented, discussing its connection to different eras political and economic contexts. The evolution of design thinking, automation and optimization highlights the importance of a holistic and interdisciplinary approach to design, integrating quantitative analysis, human-centred perspectives, and a deep understanding of societal needs.

The role of conferences in shaping the design research community and promoting an interdisciplinary approach is considered, emphasizing the importance of high-quality article reviewing standards and adequate response to societal challenges to provide theoretical support for developing innovative products, services, and systems that address challenges.

Design education, specifically problem-based learning (PBL), is considered from two perspectives. The first is focused information and communication technology

D. Marjanović (✉) · M. Štorga · S. Škec
Faculty of Mechanical Engineering and Naval Architecture, University of Zagreb, Zagreb, Croatia
e-mail: dorian@fsb.hr

D. Marjanović et al. (eds.), *Design Research: The Sociotechnical Aspects of Quality, Creativity, and Innovation*, https://doi.org/10.1007/978-3-031-50488-4_11

(ICT) tools used in design education. It discusses the dynamics and interplay between individual and collaborative use of ICT tools during a design project-based learning course, emphasizing the importance of understanding this relationship to maximize the benefits of PBL. The second perspective provides a broader, methodological approach to PBL. Highlight the benefits of PBL, such as increased student engagement, critical thinking, and real-life relevance. In addition, they emphasize the need for collaboration and the interplay between individual and team activities within PBL courses.

Several contributions tackle the challenges encountered in new product development, such as the dynamic shifts in requirements and human behaviour. These efforts specifically focus on overcoming innovation barriers and emphasize the importance of collaboration, cultural understanding, and regulatory compliance in the product development process. This is reinforced by the widespread recognition of the Human-Centered Design (HCD) methodology in recent years. HCD is acknowledged as a vital link between academic research and industrial practice, placing a strong emphasis on comprehending and catering to users' needs, behaviours, and experiences.

Ensuring the quality of design research, education, and knowledge transfer is essential for the credibility and impact of the design community. Therefore we highlight the importance of rigour and validity in research methodologies, including experimental design, data collection and analysis, and reporting findings, the essential components of high-quality research.

Each contribution provides valuable insights, perspectives, and practical approaches for educators, researchers, professionals, and developers in design. The book offers guidance for creating sustainable, responsible, and meaningful design outcomes by addressing field complexities and evolving nature.

Achieving quality, creativity, and innovation requires carefully balancing technical expertise and social factors such as collaboration, communication, and coordination. By considering the sociotechnical aspects of product development, companies can develop high-quality, innovative, and creative products that meet customer needs and drive business success. Design research will continue to evolve, consolidate, and diversify, addressing societal challenges coherently, systematically, and structured. The interdisciplinary approach, incorporating technical, social, and policy attributes, will ensure holistic research outcomes. The complex nature of new product development inspires researchers to continuously adapt to dynamic changes in requirements influenced by evolving technologies, customer attitudes, and societal expectations.